T0313740

Positioning and Location-based Analytics in 5G and Beyond

Positioning and Location-based Analytics in 5G and Beyond

Edited by
Stefania Bartoletti and Nicola Blefari Melazzi
University of Rome Tor Vergata and CNIT, Italy

IEEE PRESS
WILEY

Published by John Wiley & Sons, Inc., Hoboken, New Jersey.
Published simultaneously in Canada.

For general information on our other products and services or for technical support, please contact our Customer Care Department within the United States at (800) 762-2974, outside the United States at (317) 572-3993 or fax (317) 572-4002.

Wiley also publishes its books in a variety of electronic formats. Some content that appears in print may not be available in electronic formats. For more information about Wiley products, visit our web site at www.wiley.com.

Library of Congress Cataloging-in-Publication Data Applied for:

Hardback: 9781119911432

Cover Design: Wiley
Cover Image: © Prasit photo/Getty Images

Set in 9.5/12.5pt STIXTwoText by Straive, Chennai, India
Printed and bound by CPI Group (UK) Ltd, Croydon, CR0 4YY

C9781119911432_260923

Contents

About the Editors

Stefania Bartoletti, PhD, is an assistant professor at the University of Rome Tor Vergata, Italy, and a member of the National Inter-University Consortium for Telecommunications (CNIT). She has received research funding from the European Commission through an ERC Starting Grant, as Marie-Skłodowska Curie Global Fellow, and as coordinator of the project LOCUS.

Nicola Blefari Melazzi, PhD, is a professor at the University of Rome Tor Vergata, Italy; President of the National Inter-University Consortium for Telecommunications (CNIT); and President of the RESTART Foundation. He has received research funding from Italian Ministries, world-leading telecommunications companies, and the European Commission as coordinator of seven European projects. He has been appointed by the Italian Ministry of Research as the Italian representative to the European Smart Networks and Services Joint Undertaking.

List of Contributors

Zwi Altman
Orange Labs
Châtillon
France

Carlos S. Álvarez-Merino
Telecommunication Research Institute
(TELMA)
University of Malaga
E.T.S.I. de Telecomunicación
Málaga
Spain

Eduardo Baena
Telecommunication Research Institute
(TELMA)
University of Malaga
E.T.S.I. de Telecomunicación
Málaga
Spain

and

Telecommunication Research Institute
(TELMA)
Universidad de Málaga
Málaga
Spain

Raquel Barco
Telecommunication Research Institute
(TELMA)
University of Malaga
E.T.S.I. de Telecomunicación
Málaga
Spain

Stefania Bartoletti
Department of Electronic Engineering
and CNIT
University of Rome Tor Vergata
Rome
Italy

Giacomo Bernini
Nextworks
Pisa
Italy

Giuseppe Bianchi
Department of Electronic Engineering
and CNIT
University of Rome Tor Vergata
Rome
Italy

Hui Chen
Department of Electrical Engineering
Chalmers University of Technology
Gothenburg
Sweden

Luca Chiaraviglio
Department of Electronic Engineering
and CNIT
University of Rome Tor Vergata
Rome
Italy

Wassim B. Chikha
Orange Labs
Châtillon
France

Andrea Conti
Department of Engineering and CNIT
University of Ferrara
Ferrara
Italy

Isabel de la Bandera
Telecommunication Research Institute
(TELMA)
University of Malaga
E.T.S.I. de Telecomunicación
Málaga
Spain

Nicolò Decarli
National Research Council – Institute of
Electronics
Computer and Telecommunication
Engineering and WiLab-CNIT
Bologna
Italy

Yannis Filippas
Incelligent P.C.
Athens
Greece

Sergio Fortes
Telecommunication Research Institute
(TELMA)
University of Malaga
E.T.S.I. de Telecomunicación
Málaga
Spain

Domenico Garlisi
Department of Mathematics and
Computer Science and CNIT
University of Palermo
Palermo
Italy

Andrea Giani
Department of Engineering and CNIT
University of Ferrara
Ferrara
Italy

Domenico Giustiniano
IMDEA Networks Institute
Madrid
Spain

Carlos A. Gómez Vega
Department of Engineering and CNIT
University of Ferrara
Ferrara
Italy

Imed Hadj-Kacem
Orange Labs
Châtillon
France

Mythri Hunukumbure
Communications Research
Samsung Electronics R&D Institute UK
Staines-upon-Thames
England
United Kingdom

Sana B. Jemaa
Orange Labs
Châtillon
France

Fan Jiang
Department of Electrical Engineering
Chalmers University of Technology
Gothenburg
Sweden

Emil J. Khatib
Telecommunication Research Institute
(TELMA)
University of Malaga
E.T.S.I. de Telecomunicación
Málaga
Spain

Oluwatayo Y. Kolawole
Communications Research
Samsung Electronics R&D Institute UK
Staines-upon-Thames
England
United Kingdom

Tomasz Mach
Communications Research
Samsung Electronics R&D Institute UK
Staines-upon-Thames
England
United Kingdom

Aristotelis Margaris
Incelligent P.C.
Athens
Greece

Barbara M. Masini
National Research Council – Institute of
Electronics
Computer and Telecommunication
Engineering and WiLab-CNIT
Bologna
Italy

Marie Masson
Orange Labs
Châtillon
France

Nicola Blefari Melazzi
Department of Electronic Engineering
and CNIT
University of Rome Tor Vergata
Rome
Italy

Flavio Morselli
Department of Engineering and CNIT
University of Ferrara
Ferrara
Italy

Danilo Orlando
University "Niccolò Cusano"
Rome
Italy

Ivan Palamà
Department of Electronic Engineering
and CNIT
University of Rome Tor Vergata
Rome
Italy

Sara Modarres Razavi
Ericsson Research
Ericsson AB
Stockholm
Sweden

Athina Ropodi
Incelligent P.C.
Athens
Greece

Gurkan Solmaz
NEC Laboratories Europe
Heidelberg
Germany

Gianluca Torsoli
Department of Engineering and CNIT
University of Ferrara
Ferrara
Italy

Kostas Tsagkaris
Incelligent P.C.
Athens
Greece

Joerg Widmer
IMDEA Networks Institute
Madrid
Spain

Moe Z. Win
Laboratory for Information and Decision
Systems (LIDS)
Massachusetts Institute of Technology
Cambridge
MA
USA

Henk Wymeersch
Department of Electrical Engineering
Chalmers University of Technology
Gothenburg
Sweden

Preface

Ubiquitous 5G rollout is a main priority for Europe, as it the connectivity basis for the digital and green transformation of our economy.

Early reflection about the evolution of mobile communication networks "beyond 4G" started soon after the first deployment of a 4G commercial network in Sweden in 2010. In those days, it was already apparent that the very fast growth of mobile traffic, between 50% and 100% increase on a yearly basis, as well as the prospects to serve innovative Internet of Things (IoT) applications would drive further R&D in the mobile communication domain.

Taking note of these developments, the European Commission[1] initiated visionary EU-funded research activities already in 2012. This eventually led to the setup of the European 5G Public Private Partnership (5G PPP). The 5G PPP implemented under the European Horizon 2020 programme with about €700 Million of public support over the 2014–2020 period, the largest 5G R&D initiative in the world.

These initiatives materialize the importance of 5G networks for Europe. They are considered by the European Commission as a strategic asset for the digital society and to support the digital transformation of the industry and the public sector.[2]

The 5G PPP White Paper describing a "European Vision for the 6G Network Ecosystem"[3] highlighted that "6G is expected to play a key role in the evolution of the society towards the 2030s, as the convergence between the digital, physical and personal worlds will increasingly become a reality."

The White Paper recommended public and private R&I investment to focus on key 6G technologies, "such as programmability, integrated sensing and communication, trustworthy infrastructure, scalability and affordability, as well as AI/ML,

---

1 The views expressed in this article are those of the authors and shall not be considered as official statements of the European Commission.
2 BARANI B. & STUCKMANN P.; Leading-edge 5G Research and Innovation: An undivided commitment of Europe, 5G in Italy White Book.
3 https://5g-ppp.eu/european-vision-for-the-6g-network-ecosystem/.

microelectronics (at least in design), photonics, batteries (e.g., for mobile devices), software, and other technologies that may help to reduce the energy footprint."

Addressing the White Paper's recommendations, the Commission with EU industry set up for Horizon Europe, the new Framework Programme that started 2021, a Joint Undertaking on Smart Networks and Services, beyond 5G and toward 6G, in order to maintain Europe's technology leadership and ensure its technological sovereignty in the longer term. The Smart Networks and Services Joint Undertaking aims to foster European technological capacities as regards smart networks and services value chains. In this context, the aim is to enable European players to develop the R&I capacities for 6G technologies as a basis for future digital services in the period to 2030. The initiative also aims to foster the development of lead markets for 5G infrastructure and services in Europe. Both sets of activities (for 5G infrastructure deployment and 6G R&I) will foster the alignment of future smart networks and services with EU policy and societal needs, including competitiveness, robust supply chains, energy efficiency, privacy, ethics, and cybersecurity.[4]

In addition, the Path to the Digital Decade[5] recognizes that a sustainable digital infrastructure for connectivity is "an essential enabler for taking advantage of the benefits of digitisation, for further technological developments and for Europe's digital leadership." It, therefore, aims to achieve all populated areas covered by 5G by 2030. The SNS JU is expected to help lead markets for 5G infrastructure and services to develop in Europe.

Global standardization and spectrum harmonization are important success factors for 6G technology and focus of SNS. Both future users and suppliers need to shape key technology standards in the field of radio communications based on existing and future spectrum bands for wireless broadband, but also in next-generation network architecture to ensure the delivery of advanced service features, e.g. through the effective use of software technologies and open interfaces, while meeting energy-efficiency requirements.

For 6G, as is already the case with 5G, the European Commission supports the emergence of a single comprehensive standard ensuring interoperability, cyberse-curity, and the necessary economies of scale in an area where R&D investments are massive. While the 5G standardization process is still ongoing, it has been assessed that several hundreds of industry contributions to 3GPP originate from results of projects supported under the 5G PPP initiative, notably for what concerns (i) the Radio Access Network architecture (RAN) and (ii) the service-oriented architec-ture of the new core network.

---

4 6G SNS Draft WORK PROGRAMME 2023.
5 COM (2021) 574 final.

For 6G, the Work Programme of the SNS JU specifically includes activities designed to support the 6G standardization phase (target 2025 with first batch of 6G Study Items).

From a European perspective, it is important to continue to follow closely and support the 5G/6G standardization process so that EU policies are taken into account, maintaining strong presence of European stakeholders enlarging it to new participants, notably from the verticals, that are today little present in 3GPP debates. An inclusive standardization process is indeed a prerequisite for a global approach to standards coping with a certain divergence of market needs in the different regions. To seize the strategic opportunities for the strong industrial sectors in Europe, the standardization agenda needs to address further important use cases other than higher capacity and data rates.[6]

The Commission has been urging the standardization bodies, notably the 3GPP, and the concerned industrial actors to step-up their efforts for the rapid development of 5G standards addressing more immediate market needs, while driving a clear strategy for a 5G global standard. In line with the EU strategy targeting 5G developments in support of "vertical" industries and the wider objectives of digitizing the European industry, benefits are expected to a wide range of industrial use cases.

Several of the features in Release 17 are intended to enhance network performance for existing services and use cases, while others address new use cases and deployment options. 5G Advanced will build on Release 17, providing intelligent network solutions and covering numerous new use cases in addition to previously defined use cases and deployment options.

New Radio has supported positioning since Release 15 through the use of LTE positioning (for non-standalone deployments) and radio-access technology independent positioning (Bluetooth, wireless LAN, pressure sensors, and so on). Release 16 introduced time-based positioning methods for NR standalone deployments (multi-round-trip time (RTT), Downlink and Uplink Time Difference of Arrival), as well as an angle-of-arrival and angle-of-departure-based positioning measurements, which can be used in combination with timing-based solutions to achieve higher accuracy.

In Release-17, NR positioning is further improved for specific use cases such as factory automation by targeting 20–30 cm location accuracy for certain deployments. Release-17 also introduces further enhancements to latency reduction to enable positioning in time-critical use cases such as remote-control applications.[7]

---

6 KEMOS A., BARANI B., and STUCKMANN P. "5G Standardisation", Enjeux numériques - N.5 - mars 2019.
7 EKUDEN E. "5G evolution toward 5G advanced: An overview of 3GPP releases 17 and 18", https://www.ericsson.com/en/reports-and-papers/ericsson-technology-review/articles/5g-evolution-toward-5g-advanced.

Aside from high-positioning accuracy, industrial Internet of Things (IIoT) and automotive use cases also demand integrity protection of the location information. From a higher layer point of view, Release-17 introduces key performance indicators to indicate the reliability/integrity of the measurement report limited to the global navigation satellite system (GNSS) positioning procedure.

Precise positioning is often considered as a Satellite-based feature, with the popular use of GPS or Galileo systems. Cellular networks, sensors, local or personal area wireless technologies, and even Artificial intelligence (AI) are complementary technologies which can help provide more robust and seamless location awareness in challenging environments like indoor positioning. These use cases are getting increasingly important, as 5G intends to support demanding verticals like manufacturing processes in factories, where sub-cm positioning precision emerges as an important requirement. While positioning was not part of the essential requirements for 5G as outlined in ITU recommendation M.2083, it has become an essential feature for 5G later releases, 5G advanced, and 6G.

5G breaks technology barriers with key innovations for precise positioning already in the 3GPP Release 16 standard, with enhancements in Release 17 and in 5G Advanced. The wider bandwidths in 5G allow for finer timing resolutions. Time resolution is further improved by integrating methods for measuring, reporting, and compensating for processing delays into the radio protocol. The large number of antenna elements in massive MIMO, for mid-bands and mmWave, generates narrower radio beams which allow for finer angular resolution. With comprehensive work on time, distance, and angular precision, the most advanced versions of 5G will provide cm-level accuracy while 6G may go below that. This will be very much needed to realize the 6G vision with a massive real-time twinning of the physical and the digital worlds, which require significant progress on two aspects less explored in 5G: positioning and deterministic communications.

The 5G PPP LOCUS[8] project has led pioneering work in that domain, through improvements of the functionality of 5G infrastructures to provide accurate and ubiquitous location information as a network-native service and to derive complex features and behavioral patterns out of raw location and physical events, and expose them to applications via simple interfaces. Localization, together with analytics, and their combined provision "as a service" increase the overall value of the 5G ecosystem, allowing network operators to better manage their networks and expand the range of offered applications and services. This highly valuable work has much contributed to the production of this book, and we remain indebted to the authors for making it possible through their undivided commitment and dedication.

Bernard Barani and Achilleas Kemos
European Commission, DG CONNECT-E1

---

8 https://5g-ppp.eu/locus/.

Acknowledgments

This book was supported, in part, by the LOCUS Project through the European Union's Horizon 2020 Research and Innovation Programme under Grant 871249.

The authors wish to thank Bernard Barani and Achilleas Kemos from the European Commission, DG CONNECT-E1, for their valuable contributions to this book. Their insights and expertise have provided a valuable perspective on the subject matter and helped set the tone for the book.

List of Abbreviations

Acronyms	Definitions
3GPP	3rd Generation Partnership Project
5G	Fifth-generation
AI	Artificial intelligence
ANN	Artificial neural network
AoA	Angle-of-arrival
AoD	Angle-of-departure
API	Application programming interface
AR	Augmented reality
CAPIF	Common API framework for 3GPP northbound APIs
ETSI	European Telecommunications Standards Institute
FCC	Federal Communications Commission
GSM	Global system for mobile communications
IIoT	Industrial Internet-of-Things
IoT	Internet-of-Things
ITS	Intelligent transportation system
K-NN	K-nearest neighbors
KPI	Key performance indicator
LTE	Long-term evolution
LTE-M	Long-term evolution machine-type communication
ML	Machine learning

Acronyms	Definitions
NB-IoT	Narrowband Internet-of-Things
NWDAF	Network data analytics function
PCA	Principal component analysis
RAN	Radio access network
RAT	Radio access technology
RSSI	Received signal strength indicator
SA	System aspects
SDK	Software development kit
SVM	Support vector machine
TDoA	Time-difference-of-arrival
ToF	Time-of-flight
TSG	Technical specification group service
UE	User equipment
URLLC	Ultra-reliable low-latency communications
XR	Extended reality
V2X	Vehicle-to-everything

1

Introduction and Fundamentals

Stefania Bartoletti[1], Eduardo Baena[2], Raquel Barco[2], Giacomo Bernini[3], Nicola Blefari Melazzi[1], Hui Chen[4], Sergio Fortes[2], Domenico Giustiniano[5], Mythri Hunukumbure[6], Fan Jiang[4], Emil J. Khatib[2], Oluwatayo Kolawole[6], Aristotelis Margaris[7], Sara Modarres Razavi[8], Athina Ropodi[7], Gürkan Solmaz[9], Kostas Tsagkaris[7] and Henk Wymeersch[4]

[1]Department of Electronic Engineering and CNIT, University of Rome Tor Vergata, Rome, Italy
[2] Telecommunication Research Institute (TELMA), University of Malaga, E.T.S.I. de Telecomunicación, Málaga, Spain
[3] Nextworks, Pisa, Italy
[4]Department of Electrical Engineering, Chalmers University of Technology, Gothenburg, Sweden
[5]IMDEA Networks Institute, IMDEA, Madrid, Spain
[6]Communications Research, Samsung Electronics R&D Institute UK, Staines-upon-Thames, England, United Kingdom
[7]Incelligent P.C., Athens, Greece
[8]Ericsson Research, Ericsson AB, Stockholm, Sweden
[9]NEC Laboratories Europe, Heidelberg, Germany

1.1 Introduction and Motivation

The ever-growing demand for location- and navigation-based services has made it difficult to imagine life without the support of positioning systems. Thanks to the enhancements in 5G and other radio access technology (RAT)-independent technologies, positioning and location-based analytics are expected to have an even larger impact on many society and industry use cases today and in the future.

In cellular networks, positioning was initiated based on estimates of distance and/or direction between base stations and devices mainly to support connection establishment. The network maintained a very crude position estimate of the most recent known position of a device from global system for mobile communication (GSM) [1]. This was in order to fulfill the regulation requirements of the emergency services [2]. Since then, each generation of cellular technology has improved the level of achievable accuracy and hence enabled new applications and use cases. 3GPP developed its own positioning methods and the related localization architecture since LTE Release 9 (in 2010). In 3GPP, the term "localization" is related

Positioning and Location-based Analytics in 5G and Beyond, First Edition.
Edited by Stefania Bartoletti and Nicola Blefari Melazzi.

to the architectural and service definitions in the Service and System Aspects (SA) Technical Specification Group (TSG), and the term 'positioning' is related to the methods and implementation definitions in the Radio Access Network (RAN) TSG.

This book will cover the fundamentals of network localization, user positioning, and location-based analytics and applications in the 5G ecosystem and beyond. First, we will explore the primary verticals and relevant use cases, defining the key performance indicators (KPIs) and requirements, including those defined by the 3rd Generation Partnership Project (3GPP). The primary technologies will be described, and the foundations and signal processing approaches for accurate localization will be discussed. Architectural principles for the provision of location-based analytics will be introduced and the two main classes of such analytics will be presented: analytics for network management and analytics for verticals. Practical examples of novel solutions and applications leveraging enhanced localization and location-based analytics will be provided together with performance evaluations.

1.2 Use Cases, Verticals, and Applications

This section is organized as follows. It starts elaborating on the use cases and applications benefiting from location information, followed by the fundamentals of positioning and navigation. It then explains the fundamentals of location-based analytics and the architectural principles for the positioning solutions in 3GPP and other global initiatives.

While the first positioning use case was the location of emergency calls in prior cellular generations, currently the advances in the 5G network make it possible to target accurate and timely positioning for safety-critical intelligent transport systems (ITS) applications, such as advance warning systems and vulnerable road user protection or industrial internet of things (IIoT) scenarios. More advanced technologies such as positioning in extended reality (XR) use cases can be provisioned with 5G (B5G) and 6G.

A set of key categories of use cases, verticals, and applications that will benefit from positioning services are represented in Figure 1.1. We can identify: *emergencies, Internet of Things, IIoT, construction sites and mines, Public safety and natural disasters, ITS and Autonomous Vehicles, Commercial and transport hubs*, and *Education and Gaming*. These are detailed in the next sections.

1.2.1 Emergency Calls

The main use case that derived the positioning study in 3GPP standardization was to localize emergency calls. The Federal Communication Commission (FCC)

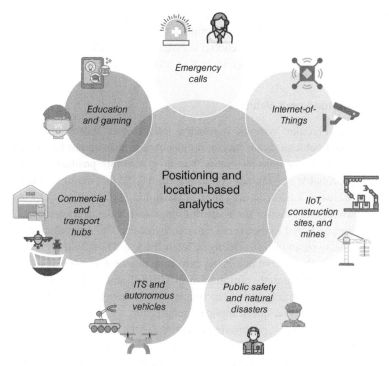

Figure 1.1 Illustration of the categories of use cases for positioning and location-based analytics.

regulatory requirements in the US mandated in 1996 that by October 2001 mobile phones calling the emergency number 911 had to be localized within 100 m for 67% of cases. In 2015, the horizontal positioning target for the user equipment (UE) was set to below 50 m for 80% of all 911 calls by April 2021. In a more recent report from FCC in 2020, the target for vertical positioning accuracy was also set, targeting a floor-level accuracy of 3 m in the 80% of indoor wireless 911 calls.

1.2.2 Public Safety and Natural Disasters

A key factor for successful emergency operations is reliable communication and real-time access to critical information. The ability to locate the victims is of course the primary goal of a rescue operation; in addition, it is critical to accurately locate first responders and/or the equipment being used throughout a rescue operation. Indeed, natural disasters and emergency events can start anywhere over large areas, and reliable and accurate fixed positioning infrastructure over the whole area is often costly and impractical. Instead, temporary and dynamic deployments for positioning can be a solution. Redundancy in the positioning technology is particularly important in these cases where, e.g. satellite signals can be obscured by vegetation or smoke. Positioning of personnel and other resources (in both horizontal and vertical dimensions) is essential for an efficient response to many kinds of emergency situations, including floods and earthquakes.

1.2.3 ITS and Autonomous Vehicles

The evolution of vehicular systems is moving toward ever more connected and fully automated vehicles. Such high level of autonomy leverages two main enablers, among others: location-awareness based on accurate positioning and sensing, and ultra-reliable low-latency communication (URLLC) among vehicles within a shared network infrastructure [3–7]. These functionalities allow vehicles to develop a shared perception of their surroundings and make decisions based on local views and expected maneuvers from nearby users. The combination of URLLC with accurate positioning and sensing leads the way toward a safer transportation system with the goal of achieving zero road deaths and a better traffic flow. Given such unprecedented combination of URLLC and high localization accuracy, 5G is the first technology that has the potential to meet some of the very stringent requirements of road safety applications [8].

Besides accuracy and latency, positioning integrity, i.e. the measure of trust that can be placed on the correctness of information supplied by a positioning system, is required for use cases such as rail and maritime, unmanned autonomous vehicles (drones), autonomous driving, and vehicle-to-everything (V2X) to minimize safety hazards, accidents, and erroneous legal decisions that involve liabilities. It is also essential for other mission-critical applications where positioning errors could cause harm, including emergency services, e-health, and many IIoT scenarios.

1.2.4 IIoT, Construction Sites, and Mines

IIoT use cases are characterized by ambitious system requirements for positioning accuracy in many verticals. For example, on the factory floor, it is important to locate assets and moving objects such as forklifts, or parts to be assembled. Similar needs exist in transportation and logistics, for example. The deployment scenario for different indoor industrial environments has a significant impact on the positioning performance in terms of both accuracy and availability of the service. The impact of the various objects that are present in a factory hall is also implicitly impacting the path loss and multipath parameters.

Tracking of tools and materials on factories, construction sites, and other industrial scenarios is of interest for increased efficiency and resource utilization. Each of the different scenarios will impose different conditions, challenges, and requirements. For instance, a construction site, in contrast to a factory, develops over time, and a supporting telecommunication infrastructure is not always available from the start. A mobile deployment of, for example, 5G base stations for positioning can quickly be put to use and be adjusted to changing needs. The conditions in the mining industry are similar in many respects: fixed infrastructure does not often exist, the environment is constantly changing, and the specific parts of the mine, where on-going work requires positioning services, vary over time.

1.2.5 Commercial and Transport Hubs

Large and open-air shopping malls, consisting of large walking and common areas with multiple shops and establishments, are a growing trend and expected to be predominant in the near future. Similarly, transport hubs, such as airports or train stations, also commonly consist of common areas together with individual shops. The network service here is highly conditioned by people's mobility and crowd aspects impacting the optimization of all network resources, together with cutting-edge applications and services. The use of ultra-dense networks in such scenarios together with the dynamic nature of the users' movement make them a very high mobility scenario, with continuous need for handovers and load balancing adaptations to cover the demand. Moreover, these are highly dense areas in terms of devices and radio equipment, where a number of heterogeneous radio access technologies are available together with potentially Internet of Things (IoT) devices. Both public and private networks (e.g. deployed by shopping malls for the customer or for the employees and logistics) may coexist. Multiple cutting-edge and very network-demanding applications are expected. XR and holographic communications, both for leisure activities in the mall as well as part of the shop activities, marketing, and customer support (e.g. augmented reality (AR)-supported navigation and recommendations). Shop logistics (e.g. goods tracking), autonomous delivery systems, environmental and security monitoring of the shopping area, flow tracking, as well as different requirements to support customers, entertainment areas, etc., make localization key to support the differentiated needs of the users and services in a cost-efficient and reliable manner. Here, enrichment information (e.g. including user position information and social-awareness [9, 10]) will be crucial to guide advanced traffic steering algorithms that will need to cope with a multi-RAT and multi-tenant scenario over heterogeneous network elements providing varied computation, energy, radio, networking, and data resources for both edge and cloud.

1.2.6 Internet-of-Things

Positioning is one of the main concepts in enabling the low-power wide-area IoT connectivity which provides a fundamental paradigm shift for people, businesses, and society. There are many applications that can benefit from IoT positioning. Some of these devices are expected to be positioned with high accuracy such as wearables, machinery control, safety monitoring, gaming gadgets, smart bicycles, medical equipment, and parking sensors. Another set of applications require position tracking while moving over a geographic area, such as assets in logistics, pets, white goods and appliances, and live stock. Moreover, there are a wide range of IoT applications where the devices have a fixed position during most of their life cycle. Some examples can be environment monitoring, soil, temperature sensors, smoke detectors, gas, water, and electricity metering, which may not have high positioning accuracy demands [1]. The advent of reduced capability (RedCap) devices, which addresses broadband IoT use cases and provides larger bandwidth than older technologies such as narrowband Internet of Things (NB-IoT) and LTE-M, will further help increasing the accuracy of IoT services.

1.2.7 Education and Gaming

Education and gaming are two very important markets that are expected to grow in the upcoming years. Education is one of the key pillars of modern society, from very early ages to university education and even mid-career training. Novel methods such as gamification and the use of XR are being explored to better engage students. These methods, along with other networked technologies (videoconferencing with tele-presence and holography, streaming, activity/ sentiment recognition), have extreme performance requirements which need to be supported by the 6G infrastructure. These techniques will have parallel applications also in the gaming market. Education and gaming will mainly require high bandwidths with low latency for communication and processing while also maintaining privacy and security to achieve a high degree of trustworthiness. These characteristics must be achieved in several different environments. While remote classes are a major novelty in the last years, with typical end-user challenges (including mobility, indoor and outdoor communications), the physical classroom will still play a very important role in the future of education. This will concentrate many users with similar extreme requirements in a small indoor area, producing very high spatial traffic densities. It will also offer some opportunities for reusing resources, such as rendered 3D objects that will be shared by all students. This proposed scenario presents significant challenges in terms of network management, including the need to balance network capacity, define active offloading strategies, and even implement in-network caching techniques. Extreme requirements are expected among the varied education applications, creating a challenge for resource allocation in an environment with multiple outdoor and indoor separate areas (e.g. classrooms, corridors).

1.3 Fundamentals of Positioning and Navigation

Wireless positioning systems estimate the location of a target node, i.e. a UE, by leveraging the communication between the UE, in unknown location, and one or multiple network access nodes, usually in known location. The estimation of the UE position relies on two main phases: (i) collection of measurements of position-dependent features performed by single or multiple nodes processing the communicated signals; and (ii) processing of such measurements employing one or multiple positioning methods, i.e. the measurements are the input of a positioning algorithm that provides the position estimate as output. There also exist advanced methods where the received signal samples are processed directly by the positioning algorithm without requiring two different steps.

1.3.1 Position-Dependent Measurements

The main measurement types used by modern positioning systems belong to the following non-exhaustive list (see Figure 1.2):

- Time-of-flight (ToF): Time taken for a signal to propagate from a target UE to a network access node (uplink) or from the node to the UE (downlink). This measurement is also referred to as time-of-arrival (ToA).

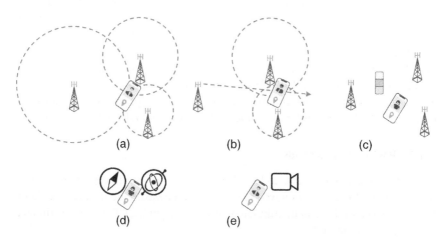

Figure 1.2 Example approaches to positioning. (a) Positioning based on ToF from several access nodes, constraining the UE to lie on the intersections of circles (or hyperbola in case of differential measurements). (b) Positioning based on a combination of time and angle measurements. (c) Positioning based on a fingerprint vector of received signal strengths. (d) Positioning based on internal sensors such as a compass and a gyroscope. (e) Positioning with a camera sensor.

- Time-difference-of-arrival (TDoA): Difference between the ToF measured by different pairs of nodes (e.g. two network access nodes paired with the same UE). This method requires synchronization of the nodes.
- Angle-of-arrival (AoA): Direction from which a signal is received at the network access node; synchronization between the devices is not required.
- Angle-of-departure (AoD): Direction from which the signal is transmitted from a network access node to the target UE. No synchronization between the devices is required.
- Received signal strength indicator (RSSI): This technique measures the intensity of signal received at the network access node.

In some applications, positioning is implemented based on internal sensors embedded in the UE, such as a compass and a gyroscope, sometimes complemented with one or more of the wireless measurements presented above. In other contexts, in addition to such measurements, computer vision analysis can be used, where vision-based systems with diverse characteristics, e.g. in the case of image sensors employed to estimate the position of the target UE. The quality of the measurement and its statistical characterization depend on the network intrinsic properties, including the nodes deployment, the signal structure, the wireless propagation conditions, and the processing itself.

In some cases, the target of a positioning system is not a UE (i.e. a device-based target that communicates with the access node), rather it is a device-free target that does not communicate with the access nodes. In such a case, the target position is inferred from signals emitted by the access nodes, backscattered by the target object, and received back by the access nodes, following a radar-like configuration. The type of measurements extracted from the received signals are similar to the device-based case (e.g. ToF, TDoA, RSSI), while the processing techniques to extract such measurements from the received signal change, as the target must be first detected while the clutter for undesired scatterers should be filtered out.

1.3.2 Positioning Methods

Once the measurements are collected by a single or multiple pairs of nodes, positioning methods are employed to process such measurements and fuse them together in a centralized or distributed manner to obtain the UE position estimate, as illustrated in Figure 1.3.

Many positioning methods leverage measurement models, i.e. models that describe the measurement collected. For example, these can include their statistical characterization, based on prior knowledge on the signal structure and wireless propagation conditions.

In some cases, a positioning system is able to collect measurements of different types and leverage them through hybrid positioning methods. The choice of the

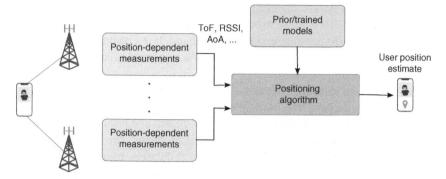

Figure 1.3 Illustration of the main blocks for estimating the user position in a cellular network.

positioning method affects the quality of the final position estimate as well as the complexity of the computation.

Navigation and tracking refer to the case where the position estimate is updated over time, considering the spatial correlation between two consecutive estimation, e.g. based on mobility models.

1.3.3 AI/ML for Positioning

Artificial intelligence (AI) and machine learning (ML) algorithms have the ability to make decisions effectively using observed data in the absence of accurate mathematical models. For example, the measurement models might be unknown to a certain degree or the solution to the problem would require prohibitively complex computations. Both these problems can be solved through AI/ML techniques, which may bring several advantages:

1. Scalability: They might be used for large-scale positioning problems when large training datasets are available.
2. Adaptability: They are flexible and can be adapted to dynamically changing environments and in the presence of multi-dimensional and heterogeneous data applications, which are common in positioning use cases.
3. Extendibility: They can be applied to fuse the information of different positioning technologies and methods, and as each positioning technology and technique has its own advantages and disadvantages, fusing them can further improve the positioning accuracy.

There are already positioning problems such as non-line-of-sight (NLOS) classification and mitigation, enhanced fingerprinting, avoiding RSSI fluctuations, trajectory learning and navigation, and fusing technologies and features that have been tackled with AI/ML algorithms with positive outcomes. The selection of which ML algorithm is suitable to be used depends on the nature and characteristics of the positioning problem and the data in hand. It is a common practice that the performance of few of these ML algorithms is compared to each other for one particular positioning problem.

Both supervised learning such as K-Nearest Neighbor (K-NN), Support Vector Machine (SVM), decision tree, random forest and artificial neural network (ANN), and unsupervised learning algorithms such as K-means and Principal Component Analysis (PCA) have been applied to positioning problems. The main distinction between the two approaches is the use of labeled datasets, meaning that supervised learning uses labeled input and output data, while an unsupervised learning algorithm does not. Moreover, reinforcement learning, deep learning, and transfer learning have also been widely attempted recently to overcome the positioning problem challenges with reasonable success [11, 12].

1.4 Fundamentals of Location-Based Analytics

Once user positioning information is available through localization, one key challenge is its application to specific services, for end-users and third parties as well as for the management of the cellular network itself. To this end, location information must be processed and combined with other information sources and variables for both user-related (e.g. people-centric) and network-related (e.g. network-centric) activities in order to generate enriched data. The resulting location-based analytics integrates positional, network, and other contextual information for mobility monitoring and to provide predictions that can be then exploited for optimizing user-centric or network-centric services through *automation, actuation*, and *decision-making* in the smart environments and the infrastructure.

In this way, Figure 1.4 illustrates a possible classification among *user-centric* and *network-centric* analytics in terms of their application. The user-centric analytics considers improving the quality of life of people through analytics services. Use cases such as smart cities, smart buildings, and human mobility can be considered in the context of user-centric analytics, where the user is the consumer of the analytics services. Differently, the network-centric analytics focus on managing the network resources, e.g. for enhancing the 5G infrastructure itself for optimal use.

The location data can be in the format of real or virtual coordinates in 2D or 3D, as well as in the format of *fingerprints* that indirectly indicate the location of the people in a given environment. In certain scenarios, ground-truth data is available for training AI/ML models, whereas in most real-world setups (*in-the-wild*) the

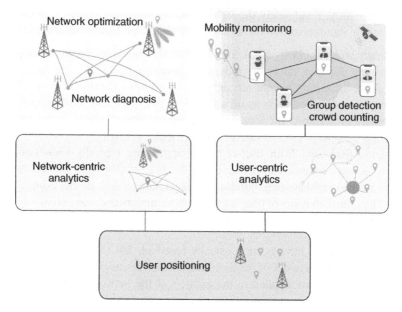

Figure 1.4 Pictorial description of the main steps for obtaining location-based network-centric (left) and user-centric (right) analytics using positioning and localization data. The user positioning can include 5G localization.

ground-truth is partially available and the system relies on the previous learning or configuration.

There have been various techniques for preprocessing of such location data sources and making it useful for the utilization by the AI/ML models. The preprocessing steps include data integration and linking, data cleaning, data sampling, normalization, data encoding, and feature selection.

User-centric (people-centric) analytics leverage different AI/ML methods for predictions and insight generation, ranging from unsupervised clustering techniques to weakly or semi-supervised models and fully supervised deep learning such as recurrent neural networks and generative adversarial networks. Real-world datasets from small indoor setups to large regional scales can be leveraged for understanding the performance of these statistical methods, their applicability, and scalability. This book presents a set of AI/ML techniques and performance evaluation metrics and describes their advantages as well as their drawbacks for real-world scenarios.

Below are two examples of real-world scenarios for people-centric analytics:

- People movement: Crowd mobility analytics applications and COVID-19 contact tracing

- Vehicular mobility: Road safety through connected vehicular system applications and their standardization including Vulnerable Road Users clustering.

The accuracy and granularity of the location data are critical for people-centric applications, e.g. for road safety or epidemic monitoring [8, 13–15]. The benefits of the above-listed applications are significant for improving the services for people using shared places, such as urban areas in cities, airports, and campus environments. Furthermore, the location-based analytics applications are useful for optimizing transportation and safety in the vehicular domain.

The insights generated from user-centric analytics are typically visualized in dashboards and consumed by either end-users (e.g. student at a university campus) or decision makers (e.g. building management services or city administration). Thus, the behaviors of people can be better understood and potentially influenced thanks to the location-based user-centric analytics.

Network-centric analytics have been classically based on UE *traces*: geolocated reports gathered by the network or the terminals themselves, as part of drive tests. The resulting analytics are applied to the analysis of the network performance for its planning, optimization, and failure management. In this context, one of the main challenges is the application of positioning information, classically 2D or 3D, to the approaches commonly followed in network management, that are typically based on events/alarms and time-series (e.g. counters, KPIs) analysis [16, 17]. Some mitigation mechanisms are based on the inclusion of the UE coordinates as general additional variables of the geolocated traces to be analyzed, or in a reference to specific points in the scenario (e.g. distance to a base station). Further processing of the data, however, helps with its integration into classical schemes, for example via the generation of statistics based on the filtering of the traces for specific areas, leading to the calculation of *contextualized indicators* [18]. The inclusion of distances or associations to other geographical information also helps to support the analysis, such as for failure management and performance forecasting. Moreover, recent advances in the development of deep learning and image-processing pave the way to the analysis based on maps constructed from geolocated measurements, e.g. for the identification of areas affected by failures.

1.5 Introduction to Architectural Principles

Context-awareness is essential for a variety of 5G vertical applications, as context depends on location information of people and objects – whether moving or stationary. Indeed, as anticipated above, location-based services are becoming more and more relevant as they impact a high number of heterogeneous use cases, verticals, and applications.

5G can substantially boost the location-based services market, as it can offer improved accuracy combined with high degree of security and location integrity when compared to 4G. Most importantly, this can be done leveraging on built-in 5G location services defined in 3GPP specifications. Indeed, as originally defined in long-term evolution (LTE) and further adopted in 5G-NR, there exist three positioning modes: (i) standalone; (ii) network-assisted; (iii) network-based. In standalone, the UE localizes itself without any aid from the network. In network-assisted, UE position is calculated by UE with assistance from the network. Finally, in network-based, UE position is calculated in the network side with measurements sent from the UE. In particular, the UE positioning in 5G can be carried out by the UE itself with assistance from the network (network-assisted) or by the network (network-based).

1.5.1 5G Architecture and Positioning

The 5G system consists of the next-generation radio access network and the 5G core network The 5G core network interacts with so-called network functions through service-based interfaces.

Location-related functionalities for any UE, including vehicle UEs, are defined within the enhanced 3GPP Location Service architecture. A control plane location service is initiated by the access and mobility management function (AMF), either on behalf of a particular UE or after request from a location services client, i.e. the network element interacting with the gateway mobile location center (GMLC) to access and process location data. The client can be anywhere in the architecture, even inside the UE.

The location service request is then sent to the location management function (LMF), i.e. the location server, which coordinates and calculates the user position. In particular, the positioning assistance information and measurements are transferred between the UE and the TRPs to and from the LMF. The sensor information is also sent from UEs to LMF, and therefore in case of high density of UEs sharing information and requesting network-based positioning, the aforementioned LMF deployment strategy would be key to meeting the demand and reducing the load on the network.

1.5.2 Location-Based Analytics Platform

Location-based analytics can be used to exploit positioning information, leveraging spatio-temporal basic analytics, as well as advanced analytics that require chaining of multiple ML and/or AI functions. Moreover, when seamlessly integrated with the 5G architecture, positioning related information/services and location data analytics can augment the standard 5G location services and functionalities and use cases, targeting an agile and on-demand exposure of analytics data and services that leverages 5G network management and orchestration features to automate their provisioning.

In practice, location-based services and related data analytics can increase the overall value of 5G if offered "as a service." This way, network operators would be able to expand their range of offered services, as they would be able to transparently combine 5G legacy services with advanced location-based services on the same end-to-end 5G cloud-native infrastructure. On the other hand, location data analytics could be exploited to improve network performance and operations and thus enable new vertical services.

As a consequence, the realization of such a location-based analytics as a service approach requires a platform capable to provide a unified approach to expose flexible and open interaction with heterogeneous localization, analytics, and ML functions. From a practical perspective, it shall enable a wide set of use cases and applications to consume location-based analytics and on-demand ML predictions and when they require for it. At the same time, it shall be also capable to decouple the exposure of the analytics services from the internal analytics data control and operation constraints. Specifically, the design and implementation of a location-based analytics platform shall follow few basic principles, which can be summarized as:

- Support innovation: The location-based analytics platform shall be based on generic and repeatable design of intelligent analytics functions and interfaces, in support of heterogeneous functionalities. These include localization enablers (which provide positioning mechanisms based on 3GPP, non-3GPP, and device-free technologies) and analytics functions (which apply descriptive, predictive, prescriptive, and diagnostic algorithms on top of geolocation data in both real and non-real time) integrated with ML/AI functionalities and capabilities. The platform shall facilitate the design and implementation of algorithms and functions and undertake all the complexity of their integration and chaining as on-demand services while hiding the internal details toward external use cases and applications.
- Implementable: The location-based analytics platform shall be based on well-defined specifications for a software architecture, at the same time following best practices in data analytics software and promoting the integration of selected relevant open source frameworks. In practice, the platform shall provide guidelines to analytics functions developers in the form of a software development kit (SDK) capable to provide a unified and common approach for functions interactions and combination into analytics services.

- Data-centric: The platform shall be designed to facilitate an easy data exchange and data consumption among heterogeneous analytics and ML/AI functions. In this context, data is key, and the platform shall be able to transparently ingest, manage, and process data feeds from 3GPP and non-3GPP sources. With the aim of providing flexibility and scalability in data management, both batch and streaming data exchange modes shall be supported, thus providing well-defined data schema on top of big data and persistence layers, query engines, and message buses/brokers.

- Dynamic, flexible, and manageable: The platform shall provide agile mechanisms to manage the complexities of heterogeneous location-based analytics functions deployment, configuration, and composition. It has to be ML-aware, enabling acceleration in the execution of pipelines and inference. To facilitate the integration with the 5G network, this has to happen on top of end-to-end 5G infrastructures, which are natively following cloud-native virtualization principles and technologies. Therefore, the location-based analytics platform shall follow an agile microservice-based design, as well as best practices for analytics functions, (cloud-native) virtualization and packaging, thus guaranteeing alignment with 5G end-to-end network management and orchestration frameworks to enable automation.

- Vertical-friendly: When providing location-based analytics as a service, the platform shall expose rich and intuitive application programming interfaces (APIs), to be consumed effortlessly by a plethora of use cases and applications. In practice, the platform northbound interface shall be in the form of an application programming interface (API) catalogue with rich metadata facilitating service discovery on the one hand, as well as subscription-based API access to hide the complexity of internal analytics functions spawning (with related dependencies and constraints). Dedicated analytics data APIs shall then provide access to the refined output of the requested analytics services or functions, based on well-known data schemas and formats.

- Impactful: High impact of the location-based analytics platform can be reached if the above principles are followed and implemented in a way to seamlessly integrate – and if necessary expand on – location-based analytics-oriented existing concepts and innovations in the ecosystem and standards (including de-facto open source standards), see e.g. Figure 1.5. Indeed, the 5G industry follows 3GPP standards to guarantee interoperability and multi-vendor cooperation. Similarly, the location-based analytics platform shall be aligned with existing work defined by 3GPP in the area of network data analytics (e.g. Network Data Analytic Function [NWDAF]) and northbound APIs toward verticals (e.g. Common API Framework [CAPIF]), as well as from European

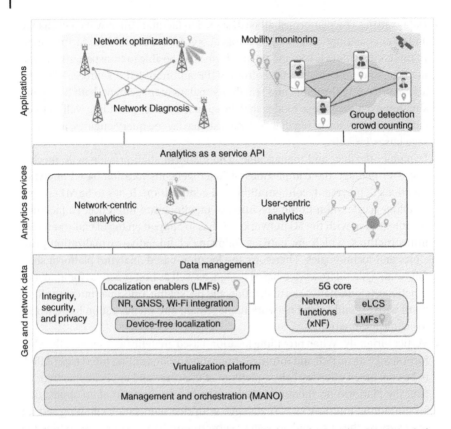

Figure 1.5 Illustration of the high-level system architecture for location-based analytics as a service.

Telecommunications Standards Institute (ETSI) for what concerns network functions virtualization (and related management and orchestration functionalities) and zero-touch service management.

1.6 Book Outline

The remainder of the book is organized in three parts. Part I presents the positioning enablers. Here, Chapter 2 presents the principles and more recent advances underpinning positioning methods; Chapter 3 focuses on 5G technology and positioning methods; while Chapter 4 introduces the main enablers and methods under discussion for positioning with beyond 5G technologies. In Chapter 5, the security, integrity, and privacy aspects of positioning and location-based analytics are presented.

Part II investigates the location-based analytics for new services in Chapter 6 and the use of analytics for network planning and management in Chapter 7.

On top of the high-level location-based analytics platform principles presented in this section, Part III with Chapters 8 and 9 go deeper by providing more insights on design and technical principles relevant for the platform implementation together with a high-level sketch of a system design.

References

1 S. M. Razavi, F. Gunnarsson, H. Rydén, A. Busin, X. Lin, X. Zhang, S. Dwivedi, I. Siomina, and R. Shreevastav. Positioning in cellular networks: Past, present, future. In *IEEE Wireless Communications and Networking Conference (WCNC)*, pages 333–338, Barcelona, Spain, April 2018.

2 TS 03.71. Location Services (LCS); Functional description; Stage 2 (Release 8), June 2004. Release 8.

3 R. Hult, G. R. Campos, E. Steinmetz, L. Hammarstrand, P. Falcone, and H. Wymeersch. Coordination of cooperative autonomous vehicles: Toward safer and more efficient road transportation. *IEEE Signal Processing Magazine*, 33(6):74–84, 2016.

4 R. Di Taranto, S. Muppirisetty, R. Raulefs, D. Slock, T. Svensson, and H. Wymeersch. Location-aware communications for 5G networks: How location information can improve scalability, latency, and robustness of 5G. *IEEE Signal Processing Magazine*, 31(6):102–112, 2014.

5 H. Wymeersch, G. Seco-Granados, G. Destino, D. Dardari, and F. Tufvesson. 5G mmWave positioning for vehicular networks. *IEEE Wireless Communications Magazine*, 24(6):80–86, 2017.

6 A. Behravan, V. Yajnanarayana, M. F. Keskin, H. Chen, D. Shrestha, T. E. Abrudan, T. Svensson, K. Schindhelm, A. Wolfgang, S. Lindberg, and H. Wymeersch. Positioning and sensing in 6G: Gaps, challenges, and opportunities. *IEEE Vehicular Technology Magazine*, 18(1):40–48, 2023.

7 TS 102 637-2. Intelligent Transport Systems (ITS); Vehicular Communications; Basic Set of Applications; Part 2: Specification of Cooperative Awareness Basic Service, March 2011. Release 1.

8 S. Bartoletti, H. Wymeersch, T. Mach, O. Brunnegard, D. Giustiniano, P. Hammarberg, M. F. Keskin, J. O. Lacruz, S. M. Razavi, J. Rönnblom, F. Tufvesson, J. Widmer, and N. B. Melazzi. Positioning and sensing for vehicular safety applications in 5G and beyond. *IEEE Communications Magazine*, 59(11):15–21, 2021.

9 S. Fortes, D. Palacios, I. Serrano, and R. Barco. Applying social event data for the management of cellular networks. *IEEE Communications Magazine*, 56(11):36–43, 2018.

10 J. Villegas, E. Baena, S. Fortes, and R. Barco. Social-aware forecasting for cellular networks metrics. *IEEE Communications Letters*, 25(6):1931–1934, 2021.

11 A. Conti, F. Morselli, Z. Liu, S. Bartoletti, S. Mazuelas, W. C. Lindsey, and M. Z. Win. Location awareness in beyond 5G networks. *IEEE Communications Magazine*, 59(11):22–27, 2021.

12 A. Nessa, B. Adhikari, F. Hussain, and X. N. Fernando. A survey of machine learning for indoor positioning. *IEEE Access*, 8:214945–214965, 2020.

13 J. A. del Peral-Rosado, G. Seco-Granados, S. Kim, and J. A. López-Salcedo. Network design for accurate vehicle localization. *IEEE Transactions on Vehicular Technology*, 68(5):4316–4327, 2019.

14 D. Giustiniano, G. Bianchi, A. Conti, S. Bartoletti, and N. B. Melazzi. 5G and beyond for contact tracing. *IEEE Communications Magazine*, 59(9):36–41, 2021.

15 S. Bartoletti, A. Conti, and M. Z. Win. Device-free counting via wideband signals. *IEEE Journal on Selected Areas in Communications*, 35(5):1163–1174, 2017.

16 S. Fortes, A. Aguilar Garcia, J. A. Fernandez-Luque, A. Garrido, and R. Barco. Context-aware self-healing: User equipment as the main source of information for small-cell indoor networks. *IEEE Transactions on Vehicular Technology*, 11(1):76–85, 2016.

17 L. Chiaraviglio, S. Rossetti, S. Saida, S. Bartoletti, and N. Blefari-Melazzi. "Pencil beamforming increases human exposure to electromagnetic fields": True or false? *IEEE Access*, 9:25158–25171, 2021.

18 S. Fortes, R. Barco, A. Aguilar-García, and P. Muñoz. Contextualized indicators for online failure diagnosis in cellular networks. *Computer Networks*, 82:96–113, 2015. Robust and Fault-Tolerant Communication Networks.

Part I

Positioning Enablers

2

Positioning Methods

Stefania Bartoletti[1], Carlos S. Álvarez-Merino[2], Raquel Barco[2], Hui Chen[3], Andrea Conti[4], Yannis Filippas[5], Domenico Giustiniano[6], Carlos A. Gómez Vega[4], Mythri Hunukumbure[7], Fan Jiang[3], Emil J. Khatib[2], Oluwatayo Y. Kolawole[7], Flavio Morselli[4], Sara Modarres Razavi[8], Athina Ropodi[5], Joerg Widmer[6], Moe Z. Win[9] and Henk Wymeersch[3]

[1]*Department of Electronic Engineering and CNIT, University of Rome Tor Vergata, Rome, Italy*
[2]*Telecommunication Research Institute (TELMA), University of Malaga, E.T.S.I. de Telecomunicación, Málaga, Spain*
[3]*Department of Electrical Engineering, Chalmers University of Technology, Gothenburg, Sweden*
[4]*Department of Engineering and CNIT, University of Ferrara, Ferrara, Italy*
[5]*Incelligent P.C., Athens, Greece*
[6]*IMDEA Networks Institute, Madrid, Spain*
[7]*Communications Research, Samsung Electronics R&D Institute UK, Staines-upon-Thames, England, United Kingdom*
[8]*Ericsson Research, Ericsson AB, Stockholm, Sweden*
[9]*Laboratory for Information and Decision Systems (LIDS), Massachusetts Institute of Technology, Cambridge, MA, USA*

The principles underpinning positioning methods have been around for many decades and leverage a plethora of technologies. Nevertheless, the emergence of new use cases in the 5G and beyond era makes it necessary to advance the existing methods and technologies to enhance positioning performance and enable location-based analytics. In this chapter, the fundamentals of positioning are described, starting from a statistical estimation perspective in Section 2.1. This is followed by an in-depth treatment of radio positioning, first focusing on device-based positioning in Section 2.2 and then on device-free positioning in Section 2.3. Finally, in Section 2.4, artificial intelligence (AI) methods for positioning are detailed.

Positioning and Location-based Analytics in 5G and Beyond, First Edition.
Edited by Stefania Bartoletti and Nicola Blefari Melazzi.
© 2024 The Institute of Electrical and Electronics Engineers, Inc. Published 2024 by John Wiley & Sons, Inc.

2.1 Positioning as Parameter Estimation

Fundamentally, the position problem is a statistical estimation problem, where an unknown vector x is inferred from a random observation z. Here, x could include the 2D or 3D location of the user equipment (UE), while z could be a vector with the samples of the received waveform or the estimated distance and/or angle from different access nodes, where such distance and angles are obtained by processing the received signal. In many cases, the model for random observation can be expressed as

$$z = h(x) + n, \tag{2.1}$$

where $h(x)$ is a known function and n is measurement noise. Therefore, the position problem can be formally stated as a maximum likelihood problem:

$$\hat{x} = \arg\max_x f_z(z; x), \tag{2.2}$$

where $f_z(z; x)$ is the likelihood function for the unknown parameter x for a given observation z. Table 2.1 lists the acronyms used in this chapter.

2.1.1 The Snapshot Positioning Problem

Under the snapshot positioning problem, the vector x, comprising at least the user 3D position and 1D clock bias, should be obtained from measurements $z_1, \ldots, z_M$. For example, we can consider the signal received from each of M connected gNBs as a separate sensor or combine them into a single measurement vector. For each measurement, we have a corresponding likelihood, $f_{z_m}(z_m|x)$. In most practical cases, different measurements come from different sources and are thus independent. This avoids the need to construct an overall likelihood $f_z(z_1, z_2, \ldots, z_M|x)$. In general, we may have access to prior information about the user state $f_x(x)$, e.g. from an outside source or earlier measurements. The solution to the positioning problem is then

$$\hat{x} = \arg\max_x f_x(x) \prod_m f_{z_m}(z_m|x). \tag{2.3}$$

Unfortunately, positioning likelihoods are rarely convex functions of x (e.g. a distance measurement constrains the user to lie on a circle or sphere, which is a non-convex set). Hence, numerical methods are used to solve problems of the form (2.3), where first an initial guess is determined from geometric reasoning (e.g. intersections of circles or lines or from the prior) or from linearized version of the problem if there are more than the minimum number of sensors. This is followed by a gradient ascent on the objective function.

Table 2.1 List of acronyms.

Acronym	Definition
5G	Fifth-generation
AI	Artificial intelligence
AP	Access point
AoA	Angle-of-arrival
AoD	Angle-of-departure
AWGN	Additive white Gaussian noise
CCRB	Constrained Cramér–Rao bound
CDF	Cumulative distribution function
CFAR	Constant false alarm rate
CRB	Cramér–Rao bound
DoA	Direction-of-arrival
DFL	Device-free localization
DNN	Deep neural network
EFIM	Equivalent Fisher information matrix
ESPRIT	Estimation of signal parameters via rotational invariant techniques
FIM	Fisher information matrix
GBT	Gradient boosted trees
gNB	Next-generation Node B
ICRB	Intrinsic Cramér–Rao bound
IIoT	Industrial Internet-of-things
IMU	Inertial measurement unit
InF-DH	Indoor factory-dense high
IOO	Indoor open office
LOS	Line-of-sight
MCRB	Misspecified Cramér–Rao bound
ML	Machine learning
MPC	Multipath component
MSE	Mean squared error
MUSIC	Multiple signal classification
NLOS	Non-line-of-sight
NR	New radio
OFDM	Orthogonal frequency division multiplexing
POC	Path overlap coefficient

(Continued)

Table 2.1 (Continued)

Acronym	Definition
RANSAC	Random sample consensus
RCS	Radar cross section
RII	Range information intensity
RMSE	Root mean squared error
RNN	Recurrent neural network
RSRP	Reference signal received power
RSRQ	Reference signal received quality
RSSI	Received signal strength indicator
SCI	Soft context information
SFI	Soft feature information
SI	Soft information
SPEB	Square position error bound
SR	Sensor radar
SRN	Sensor radar network
SVE	Single-value estimate
TDoA	Time-difference-of-arrival
TRP	Transmission reception point
UAV	Unmanned aerial vehicle
UE	User equipment
UMi	Urban Micro-cell

2.1.2 Fisher Information and Bounds

Given that positioning is a statistical estimation problem of the form (2.2), we can unleash powerful methods from Fisher information theory to predict the achievable performance. Given the vectorized observation vector $\mathbf{z}$ and all the unknown parameters $\mathbf{x}$, the root-mean-square error (RMSE) of an unbiased estimator can be lower bounded by the Cramér–Rao bound (CRB) as [1]

$$\sqrt{\mathbb{E}\{\|\hat{\mathbf{x}} - \mathbf{x}\|^2\}} \geq \sqrt{\text{CRB}(\mathbf{x})} = \sqrt{\text{trace}(\mathbf{J}^{-1}(\mathbf{x}))}, \tag{2.4}$$

where $\hat{\mathbf{x}}$ is the estimated parameter vector. The matrix $\mathbf{J}(\mathbf{x})$ is the Fisher information matrix (FIM) with elements

$$[\mathbf{J}(\mathbf{x})]_{i,j} = -\mathbb{E}\left\{ \frac{\partial \log f_{\mathbf{z}|\mathbf{x}}(\mathbf{z}|\mathbf{x})}{\partial [\mathbf{x}]_i} \frac{\partial \log f_{\mathbf{z}|\mathbf{x}}(\mathbf{z}|\mathbf{x})}{\partial [\mathbf{x}]_j} \right\}. \tag{2.5}$$

In single-stage localization, the measurement vector **z** corresponds to the samples of the received signal and if the signal model is known, the FIM can be directly calculated following (2.4). In multi-stage localization, the channel geometrical parameters η (e.g. AoAs, AoDs, delays) are estimated first and based on which the unknown states **x** (e.g. UE position, orientation, and clock offset) are calculated. In this case, the FIM of the η can be obtained first (similar as (2.5)) and the FIM of the **x** can be calculated afterward via chain rule as

$$\mathbf{J}(x) = \mathbf{T}^T \mathbf{J}(\eta) \mathbf{T} = \left(\frac{\partial \eta}{\partial x} \right)^T \mathbf{J}(\eta) \left(\frac{\partial \eta}{\partial x} \right), \tag{2.6}$$

where **T** is the Jacobian matrix getting the partial derivative of η w.r.t. **x**.

We mention a number of specialized versions of FIM and CRB, often encountered in positioning problems:

- CRB of sub-vector: In some cases the position $\boldsymbol{p} \in \mathbb{R}^3$ is part of **x**, i.e. $x = [\boldsymbol{p}^T \ \eta^T]^T$, where η is a so-called nuisance parameter (e.g. channel gain or clock bias). The CRB of $\boldsymbol{p}$ is simply $\mathrm{CRB}(\boldsymbol{p}) = \sqrt{\mathrm{trace}([\mathbf{I}^{-1}(x)]_{1:3,1:3})}$. The bound on the RMSE $\sqrt{\mathrm{CRB}(\boldsymbol{p})}$ is called the position error bound (PEB).
- Equivalent Fisher information matrix (EFIM): The EFIM of $\boldsymbol{p}$ is $\boldsymbol{J}_\mathrm{e}(\boldsymbol{p}) = ([\mathbf{I}^{-1}(x)]_{1:3,1:3})^{-1}$, which can be directly computed from $\mathbf{I}(x)$ using Schur's complement.
- Constrained and intrinsic CRB: When **x** is not part of a Euclidean space, but rather belongs to a manifold, the constrained Cramér–Rao bound (CCRB) or Intrinsic Cramér–Rao bound (ICRB) should be used. This situation occurs, for instance, when **x** contains the UE orientation.
- Posterior Cramér–Rao bound (PCRB): This bound is used when there is a prior available $p(x)$ and is common in tracking problems.
- Misspecified Cramér–Rao bound (MCRB): This bound is useful under model mismatch, when the estimator does not have access to the correct likelihood, e.g. to analyze the effect of hardware impairments.

The mentioned error bounds are often used in model-based localization methods to benchmark the performance of designed localization algorithms. In addition, they could also be used for system design and optimization, such as base station placement, resource allocation, waveform design, and so on [2]. In Section 2.1.3, we describe practical algorithms for channel geometrical parameters estimation.

2.1.3 Tracking and Location-Data Fusion

Many services, such as navigation, rely on the positioning of a moving user, requiring the position to be tracked over time. In such a case, positioning involves two separate yet related processes. The first one is the snapshot positioning problem, i.e. the estimation of a user position based on a set of current measurements, the known position of reference nodes (e.g. base stations or satellites), and possibly the information from a variety of sources (e.g. see Chapter 1). The second one is the tracking problem, i.e. recursively estimating the user position, based on the user dynamics and the snapshot measurements. Both problems are conveniently treated in a Bayesian setting, and are treated in detail in [3].

The position tracking problem can be considered as an inference problem over a dynamic system, where the positioning system infers the user state $\mathbf{x}$, which changes over time, from a sequence of noisy measurements $\mathbf{z}$ (i.e. the observations). Let $\mathbf{x}(t)$ denote the user state at time t and $f_{\mathbf{x}}(\mathbf{x}(t)|\mathbf{x}(t'))$ denote the user dynamics from time t' to time t (this information may possibly be obtained from an external sensor such as IMU). Suppose that at time t', the user has determined $f_{\mathbf{x}}(\mathbf{x}(t')|\mathbf{z}(t_1), \mathbf{z}(t_2), \ldots, \mathbf{z}(t'))$, i.e. the posterior density of the state at time t', based on all preceding measurements at times $t_1 < t_2 < t'$. Suppose the next set of measurements arrives at time t, then the user can update its posterior density by applying the following two steps, illustrated also in Figure 2.1:

1. Prediction step: Predict the user state density at time t based on the dynamics:

$$f_{\mathbf{x}}(\mathbf{x}(t)|\mathbf{z}(t_1), \mathbf{z}(t_2), \ldots, \mathbf{z}(t'))$$

$$= \int f_{\mathbf{x}}(\mathbf{x}(t), \mathbf{x}(t')|\mathbf{z}(t_1), \mathbf{z}(t_2), \ldots, \mathbf{z}(t'))d\mathbf{x}(t')$$

$$= \int f_{\mathbf{x}}(\mathbf{x}(t)|\mathbf{x}(t'))f(\mathbf{x}(t')|\mathbf{z}(t_1), \mathbf{z}(t_2), \ldots, \mathbf{z}(t'))d\mathbf{x}(t').$$

Now $f_{\mathbf{x}}(\mathbf{x}(t)|\mathbf{z}(t_1), \mathbf{z}(t_2), \ldots, \mathbf{z}(t'))$ can be interpreted as a prior density of the user state at time t, before the measurement $\mathbf{z}(t)$ is observed.

2. Correction/update step: Update the user state density to account for the current measurements

$$f_{\mathbf{x}}(\mathbf{x}(t)|\mathbf{z}(t_1), \mathbf{z}(t_2), \ldots, \mathbf{z}(t))$$

$$\propto f_{\mathbf{x}}(\mathbf{x}(t)|\mathbf{z}(t_1), \mathbf{z}(t_2), \ldots, \mathbf{z}(t'))f_{\mathbf{z}}(\mathbf{z}(t)|\mathbf{x}(t)).$$

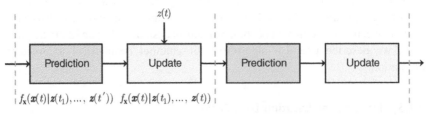

Figure 2.1 Illustration of the prediction (based on measurements up to a time $t' < t$) and update step at time t (based on measurements at time t).

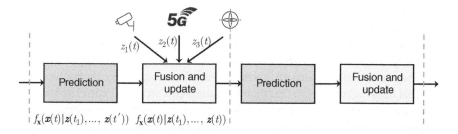

Figure 2.2 Fusion of 3 sensors (camera, 5G, and inertial measurement unit) in a tracking scenario, comprising the recursive application of prediction and update steps.

Observe that the correction step is closely related to the snapshot positioning problem, which is often a subroutine in a tracking filter.

The tracking problem considering the fusion among different sensors is visualized in Figure 2.2.

A large variety of filtering algorithms are suitable for position tracking, and their adoption usually depends on the specific pdf, as well as the requirements on complexity. The main two families of Bayesian filters are the Kalman filter and particle filters. More specifically, Kalman filters (with the extended, unscented, and cubature versions) generally have low complexity but are unable to cope with highly nonlinear models or multi-modal distributions. Particle filters can deal with highly nonlinear and non-Gaussian models, but at the cost of high computational complexity, as the number of particles grows exponentially in the state dimensionality.

2.1.3.1 Practical Aspects

A practical approach for modeling such dynamic systems is a state-space approach, where the noisy measurements $z^{(k)}$ (e.g. angle, time, power measurements) are assumed to be available at discrete times indexed by k, leading to a discrete-time formulation.

Following a state-space approach, the position $x^{(k)}$ is the state vector of the dynamic system (in many cases together with the kinematic characteristics of the user, including its speed). Such state vector is estimated and updated over time leveraging two models: (i) a *mobility model*, i.e. the state-evolution model that describes the evolution of the position with time and, (ii) a *measurement model*, which relates the noisy measurements $z^{(k)}$ to the state [4–7]. In many cases, these models are assumed as available in a probabilistic form. In other cases, approximate inference or machine learning approaches are needed.

In some cases, even for the snapshot positioning problem, the likelihood may only be partially known, or measurements may be outliers not well described by the nominal likelihood. In addition, the correlation between the different measurements may occur, in which case the likelihood cannot be factorized, and (2.3) leads to incorrect results. These issues also carry over to the tracking problem, but with several additional difficulties [8]. For instance, measurements from different sources may come at different rates or may come out of order; user dynamics may only be partially known. Moreover, sensors may themselves run internal tracking routines so that their measurements are in fact correlated across time. Finally, Bayesian tracking involves solving the complex integrals related to prediction and tracking need to be solved, either exactly or approximately. Luckily, all these problems are common among many tracking problems and powerful general-purpose solutions have been devised, e.g. the extended Kalman filter for solving the integrals.

2.2 Device-Based Radio Positioning

Device-based radio positioning is currently recognized as one of the important features in emergency, commercial, and industrial applications. This section introduces the theoretical foundations, signal processing aspects, and example results for device-based positioning.

2.2.1 Theoretical Foundations

In the case of device-based positioning, the state vector **x** corresponding to the position of the UE is estimated based on the observation vector **z** that includes multiple signal measurements from gNBs. The signal processing techniques for obtaining such measurements impact the accuracy of the position estimates. Performance benchmarks such as lower bounds for device-based positioning error are important to analyze and design localization systems, e.g. to quantify the effects of network parameters on performance and develop algorithms that can approach such bounds [9–12].

2.2.1.1 Signal Model

Consider a network consisting of N_a UEs, namely agents, and N_b gNBs, namely anchors (i.e. in known positions). The kth anchor is at $\mathbf{p}^{(k)}$. In the general case, the goal of device-based radio localization is to determine the states of UEs from inter-node and intra-node measurements as well as map information, whenever available. We focus on a simplified scenario, where the UEs do not cooperate with each other. This case essentially translates to the scenario in which $K = 1$ and $N_a = 1$, as illustrated in Figure 2.3. The state of the UE is denoted by vector x, which can include the UE's position, velocity, acceleration, orientation, and angular velocity. In a simple case, $x = [\boldsymbol{p}^T \ \boldsymbol{\eta}^T]^T$, where $\boldsymbol{p}$ is the UE position and $\boldsymbol{\eta}$ is the nuisance parameter.

Figure 2.3 Illustration of a device-based positioning configuration with $N_b = 3$ gNBs and a single UE, i.e. $N_a = 1$.

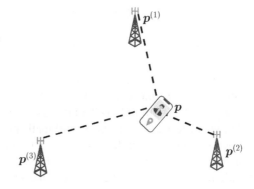

The radio signal can be transmitted by the gNB and received by the UE (i.e. downlink) or transmitted by the UE and received and processed at the gNB (i.e. uplink). The signal $r^{(k)}(t)$ is the signal received either through the kth link (i.e. between the UE and the kth gNB) after multipath propagation and can be expressed as

$$r^{(k)}(t) = \sum_{l=1}^{L^{(k)}} \alpha_l^{(k)} s(t - \tau_l^{(k)}) + n(t),\qquad(2.7)$$

where L is the number of paths (by convention $l = 1$ denotes the line-of-sight (LOS) path if present), α_l is the path gain, and τ_l is the propagation delay. The relationship between the UE position and the associated path delays is

$$\tau_l^{(k)} = \|\boldsymbol{p} - \boldsymbol{p}^{(k)}\|/c + b_l^{(k)},\qquad(2.8)$$

where $b_l^{(k)} \geq 0$ is the delay bias, and c is the propagation speed, i.e. c is the speed-of-light in free-space propagation. In particular, $b_l^{(k)} = 0$ in LOS conditions. We consider that the gNB with $k = 1, 2, \ldots, n_1$ are in LOS with the UE, while the with $k = n_1 + 1, n_1 + 2, \ldots, N_a$ are in non-line-of-sight (NLOS) with the UE. This implies that $b_1^{(k)} = 0$ for $k = 1, 2, \ldots, n_1$.

2.2.1.2 Equivalent Fisher Information Matrix
The mean squared error (MSE) of the position estimator $\hat{\boldsymbol{p}}$, according to (2.4) is lower bounded by the squared position error bound (SPEB) [13]

$$\mathbb{E}\left\{\|\hat{\boldsymbol{p}} - \boldsymbol{p}\|^2\right\} \geq \operatorname{tr}\left\{\mathbf{J}_e^{-1}(\boldsymbol{p})\right\} =: \mathcal{P}(\boldsymbol{p}),\qquad(2.9)$$

where $\mathbf{J}_e(\boldsymbol{p})$ is the EFIM as introduced in Section 2.1.2. If we consider a single-stage positioning process, i.e. the entire waveform represents the observation, then the vector $\boldsymbol{z}$ in (2.5) corresponds to the vector $\boldsymbol{r}$, i.e. the vector representation of all the received waveforms $r^{(k)}(t)$ for all the gNBs, i.e. $k = 1, 2, \ldots, N_b$. In such a case, it is demonstrated in [13] that the EFIM for the UE's position can be written as

$$\mathbf{J}_e(\mathbf{p}) = \sum_{k=1}^{n_1} \lambda_r^{(k)} \left(1 - \chi_r^{(k)}\right) \mathbf{J}_r(\phi_r^{(k)}), \tag{2.10}$$

where $\lambda_r^{(k)}$ is a nonnegative number called the ranging information intensity (RII) from the gNB k and is proportional to the signal bandwidth and the SNR for the first path of the kth link. Differently, χ_k is called path-overlap coefficient (POC) that characterizes the effect of multipath propagation for localization. Finally, $\mathbf{J}_r(\phi_r^{(k)})$ is the ranging direction matrix (RDM) which depends on $\phi_r^{(k)}$, i.e. the anchor-to-agent direction. The RDM is one-dimensional along the direction $\phi_r^{(k)}$ with unit intensity. Specifically, such an expression (2.16), which is derived using the entire waveform as observation vector, provides insights into the effect of multipath radio propagation on the position estimation.

2.2.1.3 Interpretation
When a priori knowledge is unavailable, NLOS signals do not contribute to the EFIM for the agent's position. Hence we can eliminate these NLOS signals when analyzing localization accuracy. It is also shown that, when the first path is resolvable, $\chi_k = 0$ and hence the range information intensity attains its maximum value. However, when the signal via other paths overlaps with the first one, these paths will degrade the estimation accuracy of the first path's arrival time and hence the RII. The RII of a LOS signal depends only on the first path if the first path is resolvable. In such a case, all other paths can be eliminated, and the multipath signal is equivalent to a signal with only the first path for localization.

2.2.2 Signal Processing Techniques

In the widely adopted multi-stage positioning approach, the receiver first determines the channel response from a pilot signal (e.g. by least squares estimation) and then extracts the number of paths as well as the channel parameters of each path. For example time-of-arrival (ToA)-based positioning approaches require the estimation of $\tau_1^{(k)}$ from the signal $r^{(k)}(t)$. In many scenarios, array of antennas are considered and the angle-of-arrivals (AoAs) and angle-of-departures (AoDs) are also estimated, denoted by $\theta_1^{(k)}$ and $\phi_1^{(k)}$, respectively. In such a case, the positioning problem can be solved even using a single gNB. Therefore, in the following we remove the dependency on k as we focus on a single gNB without loss of generality.

Specifically, the signal can be processed at the receiver (i.e. the UE in downlink or gNB in uplink) also in the frequency domain, which is usually the case in multi-carrier systems. In particular, under standard far-field propagation, the frequency response of the complex baseband channel is

$$\mathbf{H}(f) = \sum_{l=1}^{L} \alpha_l \mathbf{a}_R(\theta_l) \mathbf{a}_T^T(\phi_l) e^{-j2\pi f \tau_l}, \quad f \in [-W/2, W/2], \tag{2.11}$$

where W is the bandwidth, $\mathbf{a}_R(\theta_l)$ and $\mathbf{a}_T(\phi_l)$ are the array response at receiver and transmitter sides, respectively. The channel parameters are denoted by $\eta = [\eta_0, \eta_1, \ldots, \eta_{L-1}]^T$ and $\eta_l = [\alpha_l, \tau_l, \theta_l^T, \phi_l^T]^T$. The receiver first determines $\hat{\mathbf{H}}(f)$ from a pilot signal and then extracts the number of paths as well as the channel parameters of each path.

Several efficient classes of algorithms are available to solve this problem. When the number of multipaths is small (this is usually the case in mmWave systems since the path loss is significant), we can employ a compressed-sensing-based approach to reconstruct the channel matrix [14]. When the statistics of the channel geometrical parameters are available, we can apply a sparse Bayesian learning approach to perform the estimation [15]. When uniform linear/rectangular arrays and OFDM systems are employed, we can formulate the geometrical channel parameter estimation as a harmonic retrieval problem, and the classical MUltiple SIgnal Classification (MUSIC) methods and the estimation of signal parameters via rotational invariant techniques (ESPRIT) can be exploited [16]. When directional beams are used, the AoA and AoD can be read directly from the beam index, which reduces the channel parameter estimation to a one-dimensional search for the delay of each path.

Finally, it is worth pointing out that while most positioning methods rely only on the LoS path, 5G system with large antenna array and bandwidth can effectively separate the multiple signal components (MPCs), which turns the MPCs from foe to friends in the localization system.

2.2.3 Example Results of 5G-Based Positioning in IIoT Scenarios

In the industrial internet of things (IIoT) applications, the introduction of 5G comes with many advancements in terms of mobility, flexibility, reliability, and security. For example, on the factory floor, it is important to locate assets and moving objects such as forklifts. The deployment design is an important factor in network positioning performance in terms of both accuracy and availability of the service. This parameter needs even more consideration with the challenging and different IIoT environments [17].

There are many approaches on how to handle NLOS conditions suggested in the literature, see e.g. [19] for a survey. If NLOS conditions can be detected, then their effects can be mitigated by removing the positive bias or by assigning different weights to LOS respective to NLOS measurements as they are combined to estimate the UE position. NLOS measurements can even be excluded from the position estimation provided that there are sufficiently many enough reliable LOS measurements.

Figure 2.4 presents the positioning performance evaluation for the 3GPP indoor factory-dense high (InF-DH) scenario, which consists of 18 TRPs in a 120 m × 60 m

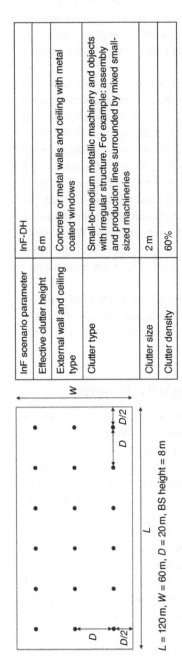

Figure 2.4 (Top) InF scenario deployment of network transmission/reception points (TRP), scenario parameters [18]. (Bottom) Positioning algorithms and their performance. Each UE is positioned using 18 measurements (one for each TRP). The x-axis of the bar-plot shows how many of these measurements that are LOS. Total number of UEs show the number of dropped UEs with LOS and the other bars show how many UEs that are located with accuracy < 1 m by each algorithm. Hence, when, e.g. the RANSAC bar is equally high as the total number of UEs, then the RANSAC algorithm locates all the UEs within 1 m from their actual position.

room, mounted on a high level and dense clutter. The scenario is implemented in a simulation environment where 800 UEs are uniformly distributed over the deployment area. Each TRP broadcast DL-PRS that the UEs perform TOA measurements on by peak detection in an estimated channel impulse response. The bandwidth and carrier frequency of the reference signal is 100 MHz and 3.5 GHz, respectively.

The UE position is estimated and the accuracy evaluated using three different methods: a least-squares (LS) method and two different variants of the RANdom SAmple Consensus (RANSAC) algorithm [20]. RANSAC is a parameter estimation method designed to cope with a large proportion of outliers in the input data. Applied to the TOA/TDOA positioning problem, this makes it well suited for environments with NLOS propagation. In Figure 2.4 (bottom) we present results both for the basic RANSAC algorithm and a modified variant with probabilistic evaluation of candidate position hypotheses (RANSAC+C). While the LS algorithm fails in presence of NLOS measurements, the RANSAC algorithms proves to be robust to situations where a large fraction of measurements are NLOS.

2.3 Device-Free Radio Localization

Device-free localization (DFL) is particularly needed in all the situations and scenarios where a person or an object needs to be tracked, without carrying dedicated devices. DFL addresses the identification and analysis of signals backscattered by single and multiple device-free targets (persons, vehicles, etc.), which are buried in the backscatter of other clutters. To this aim, one or multiple transmitters of a sensor radar network (SRN) emit wireless signals in a monitored environment. Device-free targets are detected and tracked by processing the reflected signals gathered at different receivers. DFL finds applications in several scenarios. An example of it is Industry 4.0, where mm-Wave communications enable automation and connectivity in future factories thanks to the high data rate, low latency, and low susceptibility to interference achieved by using highly directional antennas [21]. At the same time, real-time tracking of moving objects (robots, mobile machines, etc.) is an important feature for factory automation that should, however, not interfere with the normal operation of the plant.

Various technologies can support the operation of SRNs. Given the large bandwidth of mm-Wave bands, the integration of millimeter wave (mm-Wave) communications and passive localization is an extremely promising approach to design accurate, scalable, low-cost, and ubiquitous passive tracking systems. Unlike approaches employing a laser [22] which require an additional platform and resources, a wireless tracking system can be integrated with mm-Wave communication at zero cost. Another enabling technology in such scenario is ultra-wideband (UWB) [23], which allows precise resolution of multipath propagation and extremely high accuracy in time of arrival estimation [23, 24]. Sections 2.3.1–2.3.3 present DFL in different dimensions, spanning from theoretical foundations to signal processing techniques and experimentation for localization, tracking, and analytics.

2.3.1 Theoretical Foundations

The derivation of performance benchmarks for SRN in cluttered environments is challenging because the signal model is more complicated in the presence of clutter and multiple targets due to a larger number of unknown parameters.

The problem of deriving a theoretical bound for the DFL error has attracted intensive research interest. Many existing works consider a single target while clutter is modeled according to some distribution (e.g. Gaussian, compound-Gaussian, Weibull, and uniform distribution) [25, 26]. In [27], the state-of-the-art is extended, through a framework that will be able to determine the accuracy of DFL in stringent, cluttered environments.

2.3.1.1 Signal Model

Consider a multistatic SRN consisting of one transmitter at $\boldsymbol{p}_{tx}$ and multiple receivers with index set $\mathcal{R} = \{1, 2, \ldots, n_{rx}\}$, with the kth receiver at $\boldsymbol{p}_{rx}^{(k)}$, $k \in \mathcal{R}$.[1] Consider n_t device-free targets in a monitored environment, with the nth target at $\boldsymbol{p}_n$, and n_b undesired objects (i.e. walls and background objects, which are present in the environment also in the absence of targets). The presence of undesired objects in the environment has a twofold outcome: (i) the generation of background clutter, which is present in the environment also without the target; and (ii) NLOS conditions and multipath propagation for the signal reflected by the target. Background removal filters can mitigate background clutter, while they are ineffective for mitigating NLOS conditions and multipath. Specifically, the waveform gathered at the kth receiver is

$$\tilde{\mathbf{r}}^{(k)}(t) = r_t^{(k)}(t) + r_b^{(k)}(t) + \mathbf{n}^{(k)}(t), \tag{2.12}$$

where $r_t^{(k)}(t)$ is the signal component that involves the reflections from targets (affected by multipath propagation and NLOS conditions), $r_b^{(k)}(t)$ is the background signal component present also in the absence of the targets, and $\mathbf{n}^{(k)}(t)$ is

---

1 The generalization of the model to include multiple transmitters as well as the monostatic configuration is straightforward.

the additive white Gaussian noise (AWGN). In particular, the received signal component involving the reflection from targets is given by

$$r_t^{(k)}(t) = \sum_{n=1}^{n_t} \sum_{l=1}^{l_{i,n}} \sum_{m=1}^{l_{r,n}^{(k)}} \alpha_{i,n,l} \rho_n^{(k)} \alpha_{r,n,m}^{(k)} s(t - \tau_{i,n,l} - \tau_{r,n,m}^{(k)}). \tag{2.13}$$

The relationship between each target position and the associated path delays is

$$\begin{aligned} \tau_{i,n,l} &= \|\mathbf{p}_{tx} - \mathbf{p}_n\|/c + b_{i,n,l} \\ \tau_{r,n,m}^{(k)} &= \|\mathbf{p}_n - \mathbf{p}_{rx}^{(k)}\|/c + b_{r,n,m}^{(k)}, \end{aligned} \tag{2.14}$$

where $b_{i,n,l} \geq 0$ and $b_{r,n,m}^{(k)} \geq 0$ are delay biases, and c is the propagation speed, i.e. c is the speed-of-light in free-space propagation. In particular, $b_{i,n,l} = 0$ and $b_{r,n,m}^{(k)} = 0$ in LOS conditions.

Such a background signal $r_b^{(k)}(t)$ depends on the amplitudes, delays, and radar cross sections (RCSs) for each of the n_b undesired objects.[2] Then, the received waveform is processed by a background removal filter that relies on prior knowledge about the background clutter. Examples of background removal filters that are effective toward static and slowly moving clutter generators are based on successive frame difference [28].

After background removal, the received waveform can be written as $\mathbf{r}^{(k)}(t) = r_t^{(k)}(t) + \mathbf{c}^{(k)}(t)$ where $\mathbf{c}^{(k)}(t)$ describes the residual background-plus-noise disturbance. Consider $\mathbf{c}^{(k)}(t)$ as an additive white Gaussian process with zero mean and power spectral density N_{cn}.[3]

2.3.1.2 EFIM for DFL

The MSE of the position estimator $\hat{\mathbf{p}}_m$ is lower bounded by

$$\mathbb{E}\left\{\|\hat{\mathbf{p}}_m - \mathbf{p}_m\|^2\right\} \geq \mathrm{tr}\left\{\mathbf{J}_e^{-1}(\mathbf{p}_m)\right\} =: \mathcal{P}(\mathbf{p}_m). \tag{2.15}$$

Refer to $\mathcal{P}_t(\mathbf{p})$ as the total SPEB and to $\mathcal{P}(\mathbf{p}_m)$ as the individual SPEB, respectively. The square roots $\mathcal{S}_t(\mathbf{p}) = \sqrt{\mathcal{P}_t(\mathbf{p})}$ and $\mathcal{S}(\mathbf{p}_m) = \sqrt{\mathcal{P}(\mathbf{p}_m)}$ are adopted as performance metrics, together with their average values $\overline{\mathcal{S}}_t$ and $\overline{\mathcal{S}}$ with respect to $\mathbf{p}$ and $\mathbf{p}_m$.

---

2 For ease of notation, $\alpha_{b,l}^{(k)}$ represents the result of the product of path amplitude times the RCSs for the undesired objects.
3 The Gaussian model for the background plus noise residual is motivated by the fact that we are assuming a static background component generated by static or slowly moving clutter, which can be effectively removed by the background removal filter, leading to a background plus noise residual comparable to the receiver noise. The case of dynamic clutter is out of the scope of this chapter. Nevertheless, the NLOS conditions and multipath component due to the presence of undesired objects and affecting the target reflections are modeled within $r_t^{(k)}(t)$ and will be carefully taken into account in the following derivations.

To calculate the EFIM for the mth target, the Schur complement is applied again. Without loss of generality, we consider the first target and calculate $\mathbf{J_e}(\mathbf{p}_1)$. In particular,

$$
\begin{aligned}
\mathbf{J_e}(\mathbf{p}_1) = {} & \lambda_i \left(1 - \chi_i - \mu_i\right) \mathbf{J_r}(\phi_{tx}, \phi_{tx}) \\
& + \sum_{k=1}^{n_{rx}} \lambda_{ir}^{(k)} \left(1 - \chi_{ir}^{(k)} - \mu_{ir}^{(k)}\right) \left(\mathbf{J_r}(\phi_{tx}, \phi_{rx}^{(k)}) + \mathbf{J_r}(\phi_{rx}^{(k)}, \phi_{tx}) \right) \\
& + \sum_{k=1}^{n_{rx}} \lambda_r^{(k)} \left(1 - \chi_r^{(k)} - \mu_r^{(k)}\right) \mathbf{J_r}(\phi_{rx}^{(k)}, \phi_{rx}^{(k)}) \\
& - \sum_{k=1}^{n_{rx}} \sum_{j \neq k} \lambda_r^{(k)} \left(\chi_{rr}^{(j,k)} + \mu_{rr}^{(j,k)}\right) \mathbf{J_r}(\phi_{rx}^{(j)}, \phi_{rx}^{(k)}),
\end{aligned}
\tag{2.16}
$$

where the range intensity coefficients λ_i, $\lambda_{ir}^{(k)}$, and $\lambda_r^{(k)}$ have the same definitions as in the single-target case. The path overlap coefficients depend on the overlapping between incident and reflected multipaths related to the first target only, and can be calculated from eq. (58) in [27]. The multi-target overlap coefficients depend on the overlapping between inter-target multipath.

2.3.1.3 Interpretation

The EFIM for the single-path case is derived from (2.16) and results in a summation of independent contributions from all the receivers. In particular, the contribution from the kth receiver is proportional to the range intensity coefficient $\lambda_r^{(k)}$ and the range direction matrix calculated at $(\phi_{tx} + \phi_{rx}^{(k)})/2$. This result has a geometrical interpretation as in the following. Consider an ellipse with the transmitter and kth receiver as the two foci such that the target is located on the ellipse (see Figure 2.5). The angle $(\phi_{tx} + \phi_{rx}^{(k)})/2$ corresponds to the direction normal to the tangent of the ellipse at $\mathbf{p}$. Figure 2.5 shows a geometrical interpretation of the EFIM in the single-path scenario. The figure illustrates the tangent to the ellipse at $\mathbf{p}$; the arrow represents the direction of the angle $(\phi_{tx} + \phi_{rx}^{(k)})/2$, where $(\phi_{tx} + \phi_{rx}^{(k)})/2$ is the argument of the direction matrix. The figure also illustrates

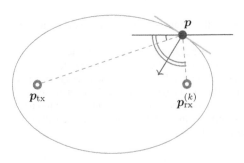

Figure 2.5 Geometrical interpretation of the EFIM in the single-path scenario [27].

the two angles ϕ_{tx} (single line angle) and $\phi_{rx}^{(k)}$ (double line angle). The kth receiver provides information only along this direction. The intuition of this result can be seen considering an ideal situation where there is no signal noise. Based on the signal reflected by the target, the kth receiver can obtain the sum of the distances between the target and the transmitter as well as the kth receiver. Therefore, the kth receiver is aware that the target is on the ellipse but is not able to locate the target. As a result, an infinitesimal change of the target position along the tangent cannot be identified by the kth receiver. This shows that the kth receiver does not provide any information of the target position along the tangent.

The EFIM in the multi-target scenario has a structure similar to the EFIM in the single target scenario and can be written as the summation of multiple terms, each one associated with a direction (defined by the direction matrix and depending on the transmitter and receiver angles with respect to each target). Moreover, each term is proportional to a range intensity coefficient and inversely proportional to an overlap coefficient. On the one hand, the range intensity coefficients and overlapping coefficient have the same interpretations as those of the single-target scenario. On the other hand, differently from the single-target scenario, the effect of inter-target multipath and associated overlapping is taken into account through the multi-target overlap coefficients.

2.3.2 Signal Processing Techniques

SRNs infer and track targets position based on the backscattered signals reflected by targets and the environment [29]. Typically, SRNs are composed of at least one transmitter and multiple spatially separated receivers that capture the backscattered signals and perform ranging to infer the target distances. In mm-Wave SRN, localization and tracking of multiple targets can be achieved using a single sensor radar (SR) equipped with multiple receiving antennas [30]. Depending on the configuration of the receiving antenna array (e.g. uniform linear array or rectangular linear array) the SR is able to infer the direction-of-arrival (DOA) of the backscattered signals both in the azimuth and elevation planes. Considering for simplicity a uniform linear array, the received signal samples can be represented in a radar data cube, where the first dimension represent the range (associated to the samples of the received signals), the second dimension represents the velocity (associated to subsequent received signals in predefined time intervals), and the third dimension represents the DOA (associated to the different receiving antennas). Such data cube serves as input to the tracking algorithm and from it the target range, radial velocity, and angle with respect to the radar can be inferred [31]. Figure 2.6 depicts an example of signal processing chain for multi-target tracking. Typically, targets are detected using constant false alarm rate (CFAR) algorithms, which determines

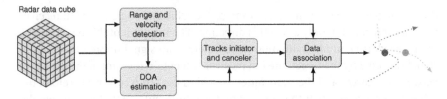

Figure 2.6 Conceptual diagram of the processing steps for multi-target tracking.

a set of measurements representing possible target ranges and velocities [32].[4] The detected ranges and velocities are then used to determine the associated DOAs. Then, the measurements (i.e. the ensemble of estimated ranges, velocities, and DOAs) are fed to a first block that takes care of initializing potential new targets track or removing old ones if no new measurements are likely to be associated with them. Lastly, data association is performed to determine which target originated a specific set of measurements and produces the estimate of the targets position [33]. Data association process predicts the current targets position based on the previous measurements and refines its prediction based on the new measurements via probabilistic approaches.

In addition to localization and tracking, counting targets (people or things) within a monitored area is an important task in emerging wireless applications (e.g. smart environments, safety, and security) [34]. Conventional approaches for device-free counting via sensor radars rely on multi-target localization or tracking [35]. Typically, this approach has a complexity that grows exponentially with the number of targets due to data association. However, data association is unnecessary when the system is only interested in crowd-centric information such as the number of targets in a given area. Such quantity can be inferred based on crowd-centric information via unsupervised learning techniques. In particular, first, a low-dimensional representation of the received waveforms is defined, based on descriptive features and principal component analysis [36]. Then, a joint probability density function (i.e. a generative model) for such representation is learned during a training phase based on previous data gathered via a measurement campaign. Finally, the generative model is used in the online phase to infer the number of targets. Figure 2.7 shows the scheme of crowd-centric counting via unsupervised learning.

2.3.3 Experimental Results on 5G-Based DFL

DFL via signals of opportunity emitted by 5G gNodeBs (gNBs) operating at mm-Wave frequencies can enable numerous applications [37]. The potential

---

4 Such measurements can be further filtered to reduce the number of false detections due to clutter residual and presence of extended targets via agglomerative clustering or morphological filtering.

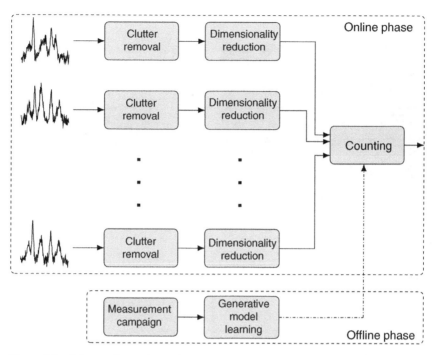

Figure 2.7 Scheme of crowd-centric counting via unsupervised learning.

use of 5G gNBs as illuminators of opportunity is due to their imminent wide availability. Moreover, the adoption of highly directional antennas and beam training schemes in mm-Wave technologies enables to efficiently capture and process the signals backscattered by device-free targets.

Figure 2.8 shows a real scenario where DFL can be performed based on DOA via 5G signals of opportunity. In particular, the transmitter of opportunity is a 5G fixed wireless access operating at 27 GHz with 100 MHz bandwidth. The lilac ellipses in this figure depict the potential beams obtained after steering the receiving antenna toward different directions with respect to the direct path. Specifically, the receiving antenna was steered toward 30° and 60° in anticlockwise direction with respect to the direct path; and the vehicle was in a direction of about 35° anticlockwise from the direct path. Note that the target is inside the area where these two potential beams overlap, and hence the receiving antenna can capture components of the back-scattered signal in both directions. The bottom-left and bottom-right figures illustrate the performed measurements with their corresponding measured spectra. The received power is significantly higher when the antenna is steered toward 30° since in this direction it can capture more of the reflections from the target. Intuitively, we can infer from the received

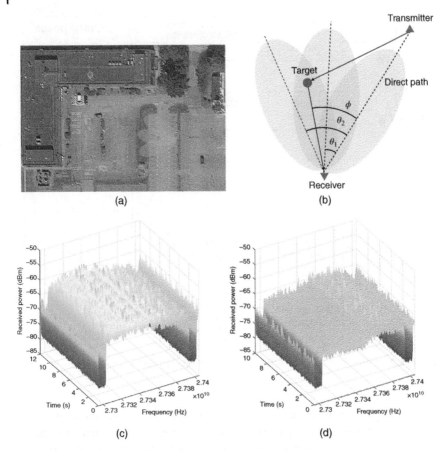

Figure 2.8 Orientation-based spectrum and pictorial direction estimation. (a) Experimental scenario, (b) antenna beams, (c) measured spectrum at 30°, and (d) measured spectrum at 60°.

power levels that the DOA is closer to 30° than to 60° anticlockwise with respect to the direct path.

2.4 AI/ML for Positioning

It is expected that AI and machine learning (ML) can significantly improve physical layer performance. In Rel-18, 3rd Generation Partnership Project (3GPP) explores the opportunities by setting up a general framework for AI/ML-related Physical layer (PHY) enhancements, including proper AI/ML modeling, evaluation methodologies, and performance requirements/testing. Positioning is one of the three use-cases that has been identified and is being studied.

On the other hand, AI and ML algorithms have been already employed for some positioning applications. The applications mostly revolve around the use of these techniques on reference signal received power (RSRP), reference signal received quality (RSRQ), and received signal strength indicator (RSSI) and other values, while there are other approaches that utilize also contextual information.

2.4.1 Fingerprinting Approach

Fingerprinting techniques are a category of positioning methods that leverage RF signal characteristics in order to identify patterns/templates that work as "fingerprints," i.e. providing a mapping between signal measurements and positions so as to be used in the identification of the UE position of a newly encountered set of signals [38]. AI and ML techniques can play an important role in these methods as they can provide an advanced framework for pattern identification and position estimation.

It should be noted that fingerprinting techniques can be based on RAT-dependent as well as RAT-independent technologies, and, alternatively, they can be built to leverage more than one technology. Several fingerprint methods have been explored leveraging different technologies, i.e. 5G, WiFi, combined 5G and WiFi, other and different AI/ML mechanisms, based on supervised, self-supervised, or unsupervised learning. A training phase takes place offline in which – after a data cleaning and pre-processing phase where necessary – the "fingerprints" are modeled. Then, this technique (i.e. the now trained model) can be utilized to infer the position of a new UE measurement (testing phase), as shown in Figure 2.9.

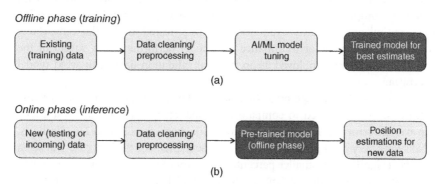

Figure 2.9 Graphical representation of the offline/training phase (a) and the online inference phase (b) processes. (a) Offline phase and (b) online phase.

Focusing on the case of supervised mechanisms, in the next paragraphs we will describe the case of WiFi measurements as input for modeling the UE position. For the training phase the availability of adequate and representative data/measurements is critical. In this particular case prior to the training phase, a simulator was used to collect the RSSI measurements from sample locations and store them in a database; however, real measurements – if available – can be used. More specifically, RSSI measurements are collected from WiFi access points (APs). As a general rule, in order to achieve a good estimation of the UE location, a number as large as possible of new radio (NR) cells and/or WiFi APs is desirable. Moreover, since the RSSI measurements suffer from shadowing and absorption attenuation, the variation of the RSSI measurements at each point does not remain constant. For this reason – and depending on the AI/ML technique – an adequate number of measurements should be acquired.

In the case of WiFi RSSI measurements, the data acquired were split into training and testing datasets, while gradient boosted trees (GBTs) and deep learning (DL) techniques, i.e. deep neural networks (DNNs), were used and compared. The first approach was chosen as a baseline model since it has been employed in similar applications [39, 40]. In fact, GBTs (and specifically the XGBoost library [41] used for optimized distributed gradient boosting) utilize multiple – but less computationally intensive – decision trees and rely on their collective decision creating an overall model by applying an iterative boosting process [42]. For best results, a hyperparameter tuning process took place, using grid search coupled with the selected squared loss function. Indicatively, these hyperparameters include the learning rate, the number of estimators (trees), and the maximum tree depth. On the other hand, DL/DNNs refer to AI methods that allow for the building of computational models (neural networks) with multiple processing layers which are able to learn representations of data with multiple levels of abstraction [43, 44]. For these neural networks with increased complexity, various dense network topologies, with different learning rates, optimizers, activations function, etc., were tested, under the TensorFlow framework [45] in order to enhance the model performance. A subset of the training dataset was employed as an internal validation dataset to avoid model overfitting.

In all cases, the average error (average Euclidean distance from the ground truth position) was used to compare performance. Comparing the best GBTs, the average error was less than 2.5 m and in the best case close to 2 m. Similarly, DNNs have very good performance – also close to 2 m – for the best case, but seem to have a significantly smaller percentage of errors for over 3–5 m. In particular, 90.8% of errors are below 3.5 m and close to 85% below 3 m, as opposed to the GBTs case yielding 83.4% and 77.5%, respectively. Lastly, it should be noted that the geospatial distribution of the errors for both cases suggested that the

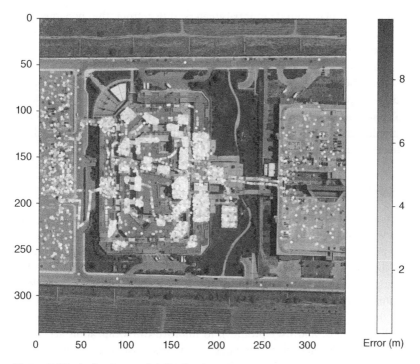

Figure 2.10 Indicative spatial distribution of measurements and error in meters in the selected simulation playground.

simulation parameters related to (a) UE mobility within the simulation playground and (b) the given WiFi AP geometry configuration may play a significant role, e.g. achieving better results in regions with high density of APs and more UE measurements, as shown in Figure 2.10.

Similarly, another DL approach was used for fingerprinting purposes. Specifically, contributing authors in the LOCUS Project investigated the use of autoencoders using a recurrent neural network (RNN)-based architecture. Autoencoders are self-supervised models, i.e. although they do not require any information/labeling for training and thus can be categorized as unsupervised, they are trained using supervised methods to produce their input as output [44]. They consist of an encoding step, which reconstructs a latent representation of an input, and a decoding step reconstructing from this representation the output, which in this case refers to the final coordinates prediction. By employing an RNN architecture and specifically a gated recurrent unit (GRU) [46] implementation, the autoencoder is able to learn from sequential aspects, i.e. temporal dependencies of the data, given the input data are ordered in time.

The simulation playground and parameters differ from the previous DL approach and results cannot be compared; however, they are similar. It was initially observed that the positioning error is close to a mean of 2 m, at most 3 m for 86% of estimates, and at most 5 m for 98% of estimates. The number of 5G cells as well as the number of measurements and the inherent bias in the mobility model used played a significant role in the autoencoder performance. Indeed, through enhanced simulation scenarios and increased number of measurements the positioning error improved and was found to be less than 1.7 m for more than 80% of the estimates.

2.4.2 Soft Information-Based Approach

As mentioned above, fingerprinting localization employs supervised and self-supervised methods in order to provide location estimates. However, unsupervised machine learning algorithms can also be exploited to significantly improve the localization accuracy with respect to classical algorithms [47–49]. In particular, soft information (SI)-based localization statistically characterizes the relation between the measurements and the device positional features (e.g. distance, direction, and velocity), as well as context information (e.g. mobility models, digital maps, and user profiles), to improve the localization accuracy. Denote with $\boldsymbol{x}^{(k)}$ the device positional state at time instant k, with $\boldsymbol{y}_i^{(k)}$ the measurement vector provided by the ith sensor where $i = 1, 2, \ldots, N$, and with $\boldsymbol{\mu}$ the contextual data. For independent measurements gathered by the N sensors (also heterogeneous), the SI is given by the posterior distribution of the device positional state as

$$f_{\boldsymbol{x}^{(k)}}(\boldsymbol{x}^{(k)} | \{\boldsymbol{y}_i^{(k)}\}_{i=1}^N; \boldsymbol{\mu}) \propto \underbrace{\prod_{i=1}^N \mathcal{L}_{\boldsymbol{y}_i^{(k)}}(\boldsymbol{x}^{(k)})}_{\text{SFI}} \underbrace{\boldsymbol{\Phi}_{\boldsymbol{\mu}}(\boldsymbol{x}^{(k)})}_{\text{SCI}}, \tag{2.17}$$

where $\mathcal{L}_{\boldsymbol{y}_i^{(k)}}(\boldsymbol{x}^{(k)})$ represents the SI associated to the measurements provided by the ith sensor, namely *SFI*, and $\boldsymbol{\Phi}_{\boldsymbol{\mu}}(\boldsymbol{x}^{(k)})$ represents the SI associated to the contextual data, namely *soft context information (SCI)*. Such posterior distribution can then be used to infer the device positional state. In complex environments, the SFI can be approximated by generative models that can be learned via unsupervised machine learning techniques including Gaussian mixture model fitting or kernel density estimation [50, 51]. Figure 2.11 depicts a pictorial representation of classical localization technique based on single-value estimations (SVEs) (left) and on SFI (right), when considering TDoA measurements in a network composed of three base stations and a single device to be localized.

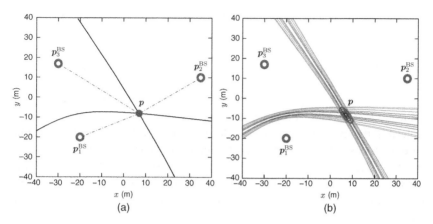

Figure 2.11 Pictorial representation of localization techniques based on single-value estimates (a) and on soft feature information (SFI) (b) for a network composed of three base stations and considering Time-difference-of-arrival (TDoA) measurements. (a) SVE-based localization and (b) SFI-based localization.

Compared to classical localization algorithms based on SVEs, SI-based localization provides significant performance improvements even in challenging wireless propagation scenarios [49]. Figure 2.12 shows on the left the cumulative density function (CDF) for the number of gNBs in LOS condition for two 3GPP standardized scenarios, namely urban micro-cellular (UMi) and indoor open office (IOO) scenarios [52]. UMi scenario is larger compared to the IOO scenario, with an area of approximately 500 m × 500 m with 19 gNBs deployed, while the IOO scenario covers an area of approximately 120 m × 60 m with 12 gNBs

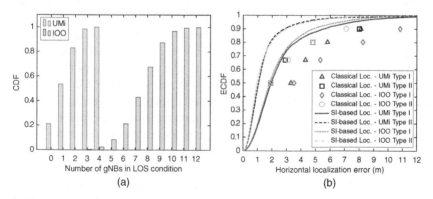

Figure 2.12 SI-based localization performance in 5G networks considering two 3GPP standardized scenarios, namely UMi and IOO. (a) CDF for the number of gNBs in LOS and (b) empirical CDF for the horizontal localization error.

deployed. It can be noticed that the UMi scenario shows a significantly lower probability of gNB in LOS condition compared to IOO scenario. Figure 2.12 shows on the right the empirical cumulative density function (ECDF) for the horizontal localization error considering both classical and SI-based localization. In particular, for classical localization performance results from 3GPP technical report are reported [52]. For each scenario 5th generation (5G) localization based on downlink TDoA is considered with different parameters of the reference signal involved in the measurement process: 50 MHz bandwidth at 2 GHz (reported in figure legend as "Type I") and 100 MHz bandwidth at 4 GHz (reported in figure legend as "Type II"). It can be noted that, at the 90th percentile, the SI-based localization improves the horizontal localization accuracy by about 2.5 m for the UMi Type I setting and by about 5 m for the UMi Type II setting. Similar performance improvements can also be observed for the IOO scenario. It can also be noted that, regardless of the scenario considered, SI provides comparable performance that is mainly determined by the reference signal bandwidth.

2.4.3 AI/ML to Mitigate Practical Impairments

For certain indoor localization scenarios, relying on the access points inside the building can become unreliable. For example, in emergency situations like building fires, some indoor access points may become dysfunctional, and methods like fingerprinting will lose validity. In such scenarios, the application of external access points (e.g. unmanned aerial vehicles unmanned aerial vehicle (UAV)s) transmitting signals into the building and the use of real-time, single-value estimation SVE methods of 5G-NR can be made to support localization of victim and emergency crew devices [53]. Here, AI/ML may be applied to develop a model to mitigate unavoidable practical impairments (e.g. UAV jitter) which can significantly affect the accuracy of the location estimates.

Supervised learning mechanism as described above can also be employed here. The jitter data was acquired from flying a UAV in the hover and throttle states at different rotor speeds in an experiment [54]. The different speeds represented different instances of wind turbulences and as shown in Figure 2.13a on the left, the severity of the jitter error increased with the speed. The acquired data was split into the training (80%) and testing (20%) datasets. During the training phase, DNN was employed to train a compensation model to predict the stable drone position from the jitter data. The predicted UAV positions obtained from the testing phase were used to estimate the errors, and it was observed the accuracy of the location estimates was comparable to their baselines in both directions (i.e. the position estimate without jitter errors). Specifically, in Figure 2.13b, the horizontal and vertical positioning errors obtained with the predicted UAV positions are 0.75 and 0.55 m at a service reliability of 95% differing only by 10 cm from the baselines.

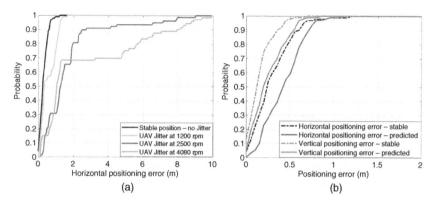

Figure 2.13 Comparison between estimated UAV positions. (a) Localization error due to Jitter data and (b) impact of NN compensation.

References

1 S. M. Kay. *Fundamentals of Statistical Signal Processing: Estimation Theory.* Prentice-Hall, Inc., 1993.

2 H. Chen, H. Sarieddeen, T. Ballal, H. Wymeersch, M.-S. Alouini, and T. Y. Al-Naffouri. A tutorial on terahertz-band localization for 6G communication systems. *IEEE Communications Surveys & Tutorials*, 24(3):1780–1815, 2022.

3 F. Gustafsson. *Statistical sensor fusion.* Studentlitteratur, 2010.

4 M. S. Arulampalam, S. Maskell, N. Gordon, and T. Clapp. A tutorial on particle filters for online nonlinear/non-Gaussian Bayesian tracking. *IEEE Transactions on Signal Processing*, 50(2):174–188, 2002.

5 A. Doucet, N. de Freitas, and N. Gordon. *Sequential Monte Carlo methods in practice.* Springer-Verlag, New York, 2001.

6 A. Doucet, S. Godsill, and C. Andrieu. On sequential Monte Carlo sampling methods for Bayesian filtering. *Statistics and Computing*, 10(3):197–208, 2000.

7 E. A. Wan and R. Van der Merwe. *The Unscented Kalman Filter.* John Wiley & Sons, Inc., New York, NY, USA, 2001. Haykin, S.

8 B. Khaleghi, A. Khamis, F. O. Karray, and S. N. Razavi. Multisensor data fusion: A review of the state-of-the-art. *Information Fusion*, 14(1):28–44, 2013.

9 J. Ziv and M. Zakai. Some lower bounds on signal parameter estimation. *IEEE Transactions on Information Theory*, 15(3):386–391, 1969.

10 D. Chazan, M. Zakai, and J. Ziv. Improved lower bounds on signal parameter estimation. *IEEE Transactions on Information Theory*, 21(1):90–93, 1975.

11 F. Gustafsson and F. Gunnarsson. Mobile positioning using wireless networks: Possibilities and fundamental limitations based on available wireless network measurements. *IEEE Signal Processing Magazine*, 22(4):41–53, 2005.

12 D. B. Jourdan, D. Dardari, and M. Z. Win. Position error bound and localization accuracy outage in dense cluttered environments. In *2006 IEEE International Conference on Ultra-Wideband*, pages 519–524, September 2006. Waltham, MA, USA.

13 M. Z. Win, Y. Shen, and W. Dai. A theoretical foundation of network localization and navigation. *Proceedings of the IEEE*, 106(7):1136–1165, 2018.

14 K. Venugopal, A. Alkhateeb, N. G. Prelcic, and R. W. Heath. Channel estimation for hybrid architecture-based wideband millimeter wave systems. *IEEE Journal on Selected Areas in Communications*, 35(9):1996–2009, 2017.

15 R. Prasad, C. R. Murthy, and B. D. Rao. Joint approximately sparse channel estimation and data detection in OFDM systems using sparse bayesian learning. *IEEE Transactions on Signal Processing*, 62(14):3591–3603, 2014.

16 M. Haardt, F. Roemer, and G. Del Galdo. Higher-order SVD-based subspace estimation to improve the parameter estimation accuracy in multidimensional harmonic retrieval problems. *IEEE Transactions on Signal Processing*, 56(7):3198–3213, 2008.

17 J. Ahlander and M. Posluk. *Deployment Strategies for High Accuracy and Availability Indoor Positioning with 5G*. Linköping University, 2020.

18 TR 38.901. 3rd Generation Partnership Project (3GPP), Technical Report Group Radio Access Network; Study on channel model for frequencies from 0.5 to 100 Ghz, March 2020. Release 16.

19 J. Khodjaev, Y. Park, and A. S. Malik. Survey of NLOS identification and error mitigation problems in UWB-based positioning algorithms for dense environments. *Springer Telecommunications*, 65:301–311, 2010.

20 R. C. Bolles and M. A. Fischler. A RANSAC-based approach to model fitting and its application to finding cylinders in range data. *International Joint Conference on Artificial Intelligence (IJCAI)*, pages 637–643, 1981.

21 B. Chen, J. Wan, L. Shu, P. Li, M. Mukherjee, and B. Yin. Smart factory of industry 4.0: Key technologies, application case, and challenges. *IEEE Access*, 6:6505–6519, 2018.

22 Y.-S. Chou and J.-S. Liu. A robotic indoor 3D mapping system using a 2D laser range finder mounted on a rotating four-bar linkage of a mobile platform. *International Journal of Advanced Robotic Systems*, 10(1):45, 2013.

23 D. Dardari, A. Conti, U. J. Ferner, A. Giorgetti, and M. Z. Win. Ranging with ultrawide bandwidth signals in multipath environments. *Proceedings of the IEEE*, 97(2):404–426, 2009.

24 S. Bartoletti, W. Dai, A. Conti, and M. Z. Win. A mathematical model for wideband ranging. *IEEE Journal on Selected Topics in Signal Processing*, 9(2):216–228, 2015.

25 S. Liu, Y. Cao, T.-S. Yeo, W. Wu, and Y. Liu. Adaptive clutter suppression in randomized stepped-frequency radar. *IEEE Transactions on Aerospace and Electronic Systems*, 57(2):1317–1333, 2021.

26 S. Fortunati, L. Sanguinetti, F. Gini, M. S. Greco, and B. Himed. Massive MIMO radar for target detection. *IEEE Transactions on Signal Processing*, 68:859–871, 2020.

27 S. Bartoletti, Z. Liu, M. Z. Win, and A. Conti. Device-free localization of multiple targets in cluttered environments. *IEEE Transactions on Aerospace and Electronic Systems*, 58(5):3906–3923, 2022.

28 B. Sobhani, T. Zwick, and M. Chiani. Target TOA association with the Hough transform in UWB radars. *IEEE Transactions on Aerospace and Electronic Systems*, 52(2):743–754, 2016.

29 M. Chiani, A. Giorgetti, and E. Paolini. Sensor radar for object tracking. *Proceedings of the IEEE*, 106(6):1022–1041, 2018.

30 B. Li, S. Wang, Z. Feng, J. Zhang, X. Cao, and C. Zhao. Fast pseudospectrum estimation for automotive massive MIMO radar. *IEEE Internet of Things Journal*, 8(20):15303–15316, 2021.

31 M. A. Richards. *Fundamentals of Radar Signal Processing*. McGraw-Hill Education, 2014.

32 M. I. Skolnik. *Radar Handbook*. McGraw-Hill, New York, NY, 3rd edition, 1970.

33 T. Fortmann, Y. Bar-Shalom, and M. Scheffe. Sonar tracking of multiple targets using joint probabilistic data association. *IEEE Journal of Oceanic Engineering*, 8(3):173–184, 1983.

34 S. Bartoletti, A. Conti, and M. Z. Win. Device-free counting via wideband signals. *IEEE Journal on Selected Areas in Communications*, 35(5):1163–1174, 2017.

35 F. Meyer, T. Kropfreiter, J. L. Williams, R. A. Lau, F. Hlawatsch, P. Braca, and M. Z. Win. Message passing algorithms for scalable multitarget tracking. *Proceedings of the IEEE*, 106(2):221–259, 2018.

36 F. Morselli, S. Bartoletti, S. Mazuelas, M. Z. Win, and A. Conti. Crowd-centric counting via unsupervised learning. In *Proceedings of IEEE Workshop on Advances in Network Localization and Navigation (ANLN), International Conference on Communications*, pages 1–6, Shanghai, China, May 2019.

37 H. Kuschel, D. Cristallini, and K. E. Olsen. Tutorial: Passive radar tutorial. *IEEE Transactions on Aerospace and Electronic Systems*, 34(2):2–19, 2019.

38 R. S. Campos and L. Lovisolo. Fingerprinting location techniques, pages 497–529. John Wiley & Sons, Ltd., 2018.

39 M. Luckner, B. Topolski, and M. Mazurek. Application of XGBoost algorithm in fingerprinting localisation task, pages 661–671. Springer International Publishing, 2017.

40 Y. Liu, Z. Deng, and L. Yin. Gradient boost decision tree fingerprint algorithm for Wi-Fi localization. In J. Sun, C. Yang, and S. Guo, editors, *China Satellite Navigation Conference (CSNC) 2018 Proceedings*, pages 501–509. Springer Singapore, Singapore, 2018.

41 XGBoost Documentation, 2022. https://xgboost.readthedocs.io/en/latest/index. html, Last Accessed: April 21, 2022.

42 T. Hastie, R. Tibshirani, and J. Friedman. *Boosting and Additive Trees*, pages 337–387. Springer New York, New York, NY, 2009. ISBN 978-0-387-84858-7. doi: https://doi.org/10.1007/978-0-387-84858-710.

43 Y. LeCun, Y. Bengio, and G. Hinton. Deep learning. *Nature*, 521:436–444, 2015.

44 I. Goodfellow, Y. Bengio, and A. Courville. *Deep Learning*. MIT Press, 2016. http://www.deeplearningbook.org.

45 Tensofllow Keras API Documentation, 2022. https://www.tensorflow.org/api_ docs/python/tf/keras/Model, Last Accessed: April 21, 2022.

46 J. Chung, C. Gulcehre, K. Cho, and Y. Bengio. Empirical evaluation of gated recurrent neural networks on sequence modeling, 2014. URL https://arxiv.org/ abs/1412.3555.

47 S. Mazuelas, A. Conti, J. C. Allen, and M. Z. Win. Soft range information for network localization. *IEEE Transactions on Signal Processing*, 66(12):3155–3168, 2018.

48 A. Conti, S. Mazuelas, S. Bartoletti, W. C. Lindsey, and M. Z. Win. Soft information for localization-of-things. *Proceedings of the IEEE*, 107(11):2240–2264, 2019.

49 A. Conti, F. Morselli, Z. Liu, S. Bartoletti, S. Mazuelas, W. C. Lindsey, and M. Z. Win. Location awareness in beyond 5G networks. *IEEE Communications Magazine*, 59(11):22–27, 2021.

50 C. M. Bishop. *Pattern Recognition and Machine Learning*. Springer-Verlag, New York, 2006.

51 J. S. Klemelä. *Smoothing of Multivariate Data: Density Estimation and Visualization*, Volume 737. John Wiley & Sons, Inc., 2009.

52 TR 38.855. 3rd Generation Partnership Project (3GPP), Technical Report Group Radio Access Network; Study on NR positioning support, March 2019. Release 16.

53 O. Kolawole and M. Hunukumbure. A drone-based 3D localization solution for emergency services. In *ICC 2022-IEEE International Conference on Communications*, pages 1–6. IEEE, 2022.

54 E. Kuantama, I. Tarca, S. Dzitac, I. Dzitac, and E. Tarca. Flight stability analysis of a symmetrically-structured quadcopter based on thrust data logger information. *Symmetry*, 10(7):291, 2018.

3

Standardization in 5G and 5G Advanced Positioning

Sara Modarres Razavi[1], Mythri Hunukumbure[2] and Domenico Giustiniano[3]

[1] *Ericsson AB, Stockholm, Sweden*
[2] *Communications Research, Samsung Electronics R&D Institute UK, Staines-upon-Thames, England, United Kingdom*
[3] *IMDEA Networks Institute, Madrid, Spain*

5G positioning is a natural component in many 5G industrial use cases and verticals, with a plethora of performance requirements in terms of accuracy, latency, availability, integrity, reliability, and power efficiency. The aim of this chapter is to explore the positioning standardization developments within 3rd Generation Partnership Project (3GPP). In Section 3.1, we first summarize what has been standardized already prior to 5G, in Sections 3.2 and 3.3 we give an overview of standardization support for 5G in different releases, and finally in Section 3.4 we conclude the chapter with some principles which can be already exploited with 5G technology, but which are not yet part of the standardization, as well as the topics which are in the 5G-Advanced positioning scope within Release 18. The continuation of this Section 3.4 is further explored in Chapter 4 in which the potential 6G positioning concepts are investigated.

3.1 Positioning Standardization Support Prior to 5G

In the framework of 3GPP, positioning technologies are generally classified as radio access technology (RAT)-dependent technologies or RAT-independent technologies. For many use-cases, it may also be worth considering a combination of RAT-dependent and RAT-independent positioning techniques. RAT-independent positioning leverages the fact that the user equipment (UE) has several interfaces beyond the cellular transceiver that could be accessed and exploited such as global navigation satellite system (GNSS), WiFi, and Bluetooth. RAT-independent positioning methods have no impact on the physical layer design, although there are standardization supports to report measurements

Positioning and Location-based Analytics in 5G and Beyond, First Edition.
Edited by Stefania Bartoletti and Nicola Blefari Melazzi.
© 2024 The Institute of Electrical and Electronics Engineers, Inc. Published 2024 by John Wiley & Sons, Inc.

and receive assistance information from the cellular network in respect to these technologies. 3GPP developed its own positioning methods and the related localization architecture since LTE Release 9 (in 2010). In 3GPP, the term 'localization' is related to the architectural and service definitions in the Service and System Aspects (SA) Technical Specification Group (TSG) and the term 'positioning' is related to the methods and implementation definitions in the Radio Access Network (RAN) TSG.

To explore the positioning support prior to the 5G time frame, in this section we provide a comprehensive list of methods that are also supported by 5G but have been introduced in the 3G, 4G time frame while considering the RAT-dependent/independent categorization. Table 3.1 lists the acronyms used in this Chapter.

3.1.1 GNSS and Real-Time Kinematics (RTK) GNSS Positioning

GNSS refers to a constellation of satellites providing signals from space that transmit positioning and timing data. With this data, a UE is able to determine its own location based on the concept of trilateration. Global positioning system (GPS) was developed by the US Department of Defense for military use back in the 1970s and is one of the most widely used GNSS and a positioning method supported by 3GPP on 3G and early 4G devices. The support for provisioning of GNSS at accuracies of a few meters has been considered since the initial support of positioning in 3GPP long-term evolution (LTE).

The time of arrival measured by correlating each received satellite signal with a replica which is based on the same pseudo-random code as the satellite signal is called code phase measurement. This enables a time of arrival estimation at a fraction of the length of a code symbol, corresponding to even less than a meter. The carrier phase measurement, in contrast, is obtained when tracking the carrier frequency component of the satellite signal. It can be estimated at an accuracy of a few millimeters.

The real time kinematics (RTK) GNSS support in Rel-15 is the major approach in providing high accuracy positioning for outdoor 5G positioning use-cases. The RTK GNSS exploits the carrier phase of the GNSS signal rather than only the code phase. The user equipment (UE) is able to send both its code and carrier phase measurements for each satellite to the location server in the network, and then the location server, with the help of RTK network correction data, can provide a more accurate RTK GNSS position to the UE. Another approach is that the network provides the UEs with the RTK correction data information directly via unicast or broadcast, which the UE uses to accurately estimate its own RTK GNSS position. In principle, RTK GNSS can provide positioning accuracies down to a few millimeters (centimeters levels with lower cost hardware), albeit some quite long initialization time is needed.

Table 3.1 List of acronyms.

Acronym	Definition
3GPP	3rd Generation Partnership Project
AF	Application function
AMF	Access and mobility function
AP	Access point
A-AoA	Azimuth angle-of-arrival
AI	Artificial intelligence
AoD	Angle-of-departure
CID	Cell ID
CoO	Cell-of-origin
CSI-RS	Channel state information – reference signal
DL	Down-link
DMRS	De-modulation reference signal
eCID	enhanced Cell ID
eMBB	enhanced Mobile broadband
eNB	enhanced Node B
eLCS	enhanced Localization services
FCC	Federal communications commission
FR1	Frequency range 1
FR2	Frequency range 2
FTM	Fine time measurements
GMLC	Gateway mobile location center
GNSS	Global navigation satellite systems
GPS	Global positioning system
IMU	Inertial measurement unit
INS	Inertial navigation satellite system
LCS	Localization services
LMF	Location management function
LMU	Location management unit
LTE-M	Long-term evolution – machine type
LPI	Location privacy indication
LPP	LTE positioning protocol
LRF	Location retrieval function

(Continued)

Table 3.1 (Continued)

Acronym	Definition
LOS	Line-of-sight
ML	Machine learning
MBS	Metropolitan beacon systems
mMTC	massive Machine-type communications
MO-LR	Mobile originated location request
MT-LR	Mobile terminated location request
N3IWF	Non 3GPP inter-working function
NBIoT	Narrow band Internet-of-things
NEF	Network exposure function
NF	Network function
NG-RAN	Next-generation radio access network
NI-LR	Network-induced location request
NRPPa	New radio positioning protocol a
OFDM	Orthogonal frequency division multiplexing
OTDoA	Observed time difference of arrival
PAPR	Peak to average power ratio
PFL	Positioning frequency layers
PLMN	Public land mobile network
PRB	Physical resource block
PRS	Positioning reference signal
PUSCH	Physical uplink shared channel
RAN	Radio access network
RAT	Radio access technology
RedCap	Reduced capability
RRC	Radio resource control
RRM	Radio resource management
RSTD	Received signal time difference
RSRP	Reference signal received power
RSS	Received signal strength
RTT	Round trip time
Rx	Receiver
QoS	Quality of service
SBA	Service-based architecture

Table 3.1 (Continued)

Acronym	Definition
SSB	Single side band
SRS	Sounding reference signal
TA	Timing advance
TBS	Terrestrial beacon systems
ToA	Time of arrival
TRP	Transmit receive point
TNGF	Trusted non-3GPP function
Tx	Transmitter
UDM	Unified data management function
UE	User equipment
UL	Uplink
UTDoA	Uplink time difference of arrival
URLLC	Ultra reliable and low latency communications
UWB	Ultra wide band
V2X	Vehicle to everything
XR	Extended reality
Z-AoA	Zenith angle-of-arrival

Both of these methods are categorized as RAT-independent technologies and are used for outdoor positioning. Due to the limited satellite visibility in indoor environments, complementary positioning solutions are needed to cover indoor and urban canyons scenarios.

3.1.2 WiFi/Bluetooth-Based Positioning

While WiFi and Bluetooth technologies have their standalone positioning methods, the UE ranging measurements together with the location of the corresponding WiFi or Bluetooth access points (APs) can also be used in the cellular network for positioning purposes. 3GPP added this support in Rel-13 for LTE, and now the support is also available for NR. As these types of APs are more widely available indoors, they are considered as alternative localization solutions for indoor environments.

There are some similarities and differences concerning WiFi and Bluetooth measurements. The UE can measure both the WiFi and Bluetooth APs identifiers.

Besides that, the UE can optionally measure the received signal strength (RSS) of both WiFi and Bluetooth APs, but round trip time (RTT) of only WiFi APs. The procedure for WiFi positioning can be network-assisted or network-based. For instance, network-assisted could be used to receive the locations of WiFi APs. There is no direct communication between the 3GPP location server and WiFi APs or Bluetooth APs devices specified in 3GPP standards.

3.1.3 Terrestrial Beacon System

A terrestrial beacon system (TBS) consists of a network of ground-based transmitters, broadcasting signals only for positioning purpose. The TBS positioning signals can be similar to positioning reference signal (PRS) (i.e. PRS-based TBS) or GNSS-like signals (e.g. GPS, GLONASS, Galileo signals). The Metropolitan beacon system (MBS) is one example of the latter category. The messaging support for TBS was specified during Rel-13 as a solution to fulfill the FCC emergency localization requirements.

3.1.4 Sensor Positioning

Some sensor measurements in the UE can provide useful information in terms of positioning relative to some reference point. In Rel-13, the support for barometric pressure sensors provided a solution for UE's vertical position estimation. The measurements in this support are always reported based on a known pressure of a reference vertical position.

Later on in Rel-15, the signaling support for inertial measurement units (IMUs), which are now being widely adopted in UEs, has been also standardized. inertial measurement unit (IMU), which is also referred as inertial navigation system (INS), is based on motion sensors (accelerometers), rotation sensors (gyroscopes), and occasionally magnetic sensors (magnetometers) that continuously calculate via dead reckoning, orientation, and velocity (direction and speed of movement) of the UE. This information in combination with a reference point information reporting message from the UE can help the network to provide further improved positioning accuracy and tracking capabilities for moving UEs.

Sensor positioning is also categorized as RAT-independent positioning technology, and one should note that it only provides relative position information and therefore cannot be used as a standalone positioning technology.

3.1.5 RAT-Dependent Positioning Prior to 5G

In general, the RAT-dependent positioning technologies leverage a set of measurements that are collected either at the UE or at the radio network node (i.e. eNodeB (eNB) in LTE or gNodeB (gNB) in NR) which provide further understanding in terms of the distance, the angle, or some characteristics of the UE's location. The position estimation can be done either at the network location server (i.e. enhanced mobile location server (ESMLC) in LTE or location management function (LMF) in NR) or at the UE depending on how and where these measurements are collected. In this section, the RAT-dependent positioning methods which were introduced prior to 5G are summarized, while all of them have been enhanced to 5G and beyond.

3.1.5.1 Enhanced CID (eCID)

The eCID positioning method was first introduced in LTE Rel-9. Prior to this, the cell identity (CID) or cell of origin (CoO) method utilized information of a connected cell in the radio resource control (RRC) connected mode to position the UE to the coarse granularity of the cell area. However, the regulatory requirements for positioning around that time meant that greater accuracies in UE positioning were needed, which prompted the development of eCID.

With eCID, the inclusion of timing- or angle-based information helps to achieve positioning accuracies which are better than those of the CID method. The timing information can be derived from the timing advance (TA) or the RTT. In the uplink, the active UEs are located at different distances to the serving gNB, and without a TA, the uplink data in the shared channel (PUSCH) would arrive at the gNB at different times. Thus, the gNB derives a TA from the timing differences in the first transmissions and receptions in the Random Access process and reports this to the UE. This TA can be used to estimate the distance of the UE from the serving gNB. In the same manner, the UE can measure the timing difference for the downlink and can report this to the gNB. Combining these two timing differences, the RTT can be derived. The value of RTT/2 can be used to determine the position of the UE. Just as with the TA, the RTT with respect to only the serving gNB produces positioning results that lie on the circumference of a circle.

The positioning accuracy can be enhanced if timing information from multiple neighbor gNBs can be extracted. In this case, the UE position can be estimated as the intersection point of the respective circles. If the eCID method uses the TA, it falls within the uplink-based positioning solutions and if it used the RTT it belongs to both the uplink- and downlink-based solutions.

The eCID method can also use angle information. This is relevant when antenna arrays are used at the gNB and the phase difference between received uplink signals in each antenna element can be estimated. This allows to determine the angle-of-arrival (AoA) for the UE signal with the help of sounding reference signal (SRS) or demodulation reference signal (DMRS). If the eCID method uses the AoA, it becomes an uplink-based solution.

3.1.5.2 Observed Time-Difference-of-Arrival (OTDoA)

The Observed time-difference-of-arrival (OTDoA) positioning method was also introduced in LTE Rel-9 and enhanced in Rel-13 and 14. In OTDoA, the UE measures the time-of-arrival (ToA) of PRSs from multiple base stations and computes the relative difference with respect to a reference cell and reports them to the network as received signal time difference (RSTD). The position is estimated by multilateration techniques using the reported RSTD from the UE at the ESMLC.

3.1.5.3 Uplink Time-Difference-of-Arrival (UTDoA)

Similar to OTDoA, Uplink time-difference-of-arrival (UTDoA) is a multilateration positioning method, which uses the uplink signal (i.e. SRS). Both the ToA measurement and the multilateration computation is done at the network side. The support of UTDoA which was introduced in Rel-11 is based on having location management units (LMU) in the network architecture to measure the uplink signals sent from the UEs. This extra hardware requirement was one of the main reasons why the UTDoA never became commercially available in any production mobile networks after the standardization.

3.1.6 Internet of Things (IoT) Positioning

Positioning can also provide added value for IoT applications. A significant effort was carried out during Rel-13 to develop cellular systems that provide low power wide area IoT connectivity. Two prominent 3GPP IoT technologies were LTE-M Category-M1 (Cat-M1) and Narrowband-IoT (NB-IoT). Cat-M1 is based on LTE and has a minimum system bandwidth of 1.4 MHz (equivalent to 6 LTE physical resource blocks (PRB)), and NB-IoT belongs to NR and uses a system bandwidth of 180 kHz (equivalent to 1 PRB) to provide more flexible deployment. While both technologies had some basic positioning support in Rel-13, due to the importance of positioning features for IoT applications, many positioning enhancements mainly in the area of OTDoA and some for eCID were added in Rel-14 for both technologies tailored to their properties [1].

3.1.7 Other Non-3GPP Technologies

There are also other positioning technologies which are being used for different use-cases that have not been standardized in 3GPP. This section includes a list of non-3GPP technologies which are widely used in industry, enterprises, and sometimes even in cellular networks.

3.1.7.1 UWB

Ultra-wideband (UWB) technology uses very short (few nanoseconds), high bandwidth pulses for communication and location. UWB signals are robust to fading and provide fine delay resolution, which is desirable for communication and localization [2–4]. This technology has multiple advantages such as centimeter-level ranging precision [5], good obstacle-penetration capabilities [6] and multipath mitigation in dense scenarios [7]. In communications, this implies a very high throughput. Regarding localization, it provides a high ranging accuracy thanks to the high bandwidth, even in environments with harsh propagation characteristics [5]. To measure RTT, the devices transmit a short pulse and receive a response after a predefined time interval. This allows the devices to measure the RTT between them, without the need of clock synchronization. This RTT can then be translated into a very precise range estimate, and, with trilateration, into a location estimation.

UWB signals are centered at 3.5 GHz with a bandwidth higher than 500 MHz. The latest market trends show that UWB has the potential to become a de-facto standard for positioning and will eventually be addressed by 3GPP standards [8]. Accordingly, some smartphones have integrated UWB chipsets in the recent years [9]. The Federal Communication Commission (FCC) authorized the unlicensed use of UWB in the range of 3.1–10.6 GHz [10]. As a drawback, to achieve the short pulse width the UWB device has a high energy consumption.

3.1.7.2 Fingerprinting

Fingerprinting is not a technology but rather a method which can provide high-quality position estimates. In this method, the UE's position is found by mapping obtained RF measurements from the UE onto an RF prediction map based on site survey results. This approach can be generalized by mapping further information aside from RF measurements, such as CID, TA, AoA, and RSTD of the UE onto an automated built up database. This network-based method relies on signaling support for collecting the necessary measurements and UE locations to build the required maps. However, as the whole method would be covered at the network side, without any further measurement support from the UE, no 3GPP standardization support is required for this method.

3.2 5G Positioning

5G new radio (NR) was initially introduced as a non-standalone extension to 4G [11]. In 3GPP Rel-15, 5G device positioning was enabled by an overlay 4G network, providing 4G positioning reference signals to measure on. To leverage the multiple sensors available in today's devices, the support for technologies independent of 3GPP RAT, such as GNSS, Bluetooth, barometric pressure, WiFi signal strength, and inertial sensors, was naturally extended from 4G to 5G. In Rel-16, LTE positioning was extended to accommodate enablers of 5G such as wide-band signals,

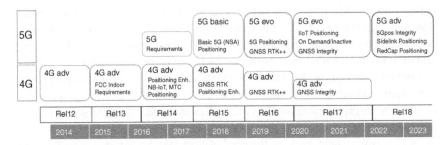

Figure 3.1 3GPP 4G and 5G positioning timeline.

higher frequencies, multiple antennas, low latency, and flexible architecture. A few key enablers for precise positioning in 5G include millimeter-wave frequency bands, which enable wide-band signals, beamforming, and precise angle estimation with multiple antennas [12]. Figure 3.1 provides the 4G (LTE) and 5G (NR) positioning timeline in 3GPP. This figure shows how the topic of positioning has continuously evolved in 3GPP from 2014 until today.

3.2.1 5G Localization Architecture

The Service-based architecture (SBA) is a main tenant in 5G and the localization architecture also follows this concept. Basically, Service based architecture (SBA) moves away from the fixed point-to-point connections between the core network nodes in 4G. In 5G, the nodes are re-defined as 'functions' based on the service(s) they provide, and the control plane functions have the flexibility to inter-connect through service-based interfaces. This flexibility is essential in 5G to support different service requirements (of eMBB, URLLC, and mMTC) for various classes of users, which may be configured as different network slices, for example. The functional architecture for 5G localization is depicted in Figure 3.2. On the right of the Figure 3.2 are the location applications (Apps) which request localization inputs from the 5G network. An application can be an external localization service (LCS) client or an application function (AF) internal to the 5G network.

In the middle of the diagram are the core network functions, handling the localization requests. An external LCS client will communicate with the gateway mobile location center (GMLC) to make the localization request and to receive the results. An internal AF will communicate through the network exposure function (NEF) with the GMLC and other functions. The location retrieval function (LRF) can be co-located with the GMLC and is only used for retrieving or validating location information for a UE which has initiated an IP and multimedia subsystem (IMS) emergency session. If the target UE is roaming, the home GMLC (H-GMLC) which receives the request will forward this to the relevant visiting

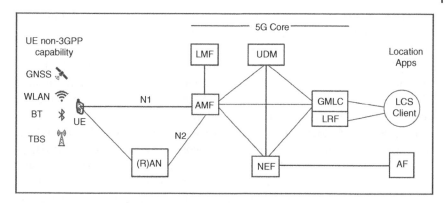

Figure 3.2 Functional architecture for 5G localization [13].

GMLC (V-GMLC). The GMLC (or V-GMLC in roaming) will forward a location request to the access and mobility function (AMF).

The AMF is responsible for managing the positioning of a target UE for both commercial and emergency/law enforcement applications. A location request originating from a UE will be first directed to the AMF. The AMF will access the RAN via the N2 reference point and the target UE via the N1 reference point. It will select a suitable LMF to execute the location request through orchestrating the RAN. The LMF also handles the overall co-ordination and scheduling of resources required for positioning of a target UE and verifies the location result(s) coming from the positioning exercise. The unified data manager (UDM) contains the localization-related privacy profile and the routing information (like for roaming) for the UE.

To the left of the core network in Figure 3.2 are the RAN and the target UE for localization. Multiple 3GPP positioning methods, both RAT-dependent and RAT-independent, can be used to derive the localization result. The selection of the positioning method can be done by the LMF, by first referring to the UE capabilities which are stored at the AMF and also considering the quality of service (QoS) requirements for the localization request. In the RAN notation, the R is in brackets as (R)AN as this Access Network can be 5G or LTE-Advanced (together known as next generation-radio access network (NG-RAN)) or even non-3GPP WiFi, when this is supported through inter-working functionality. In case WiFi is used through non-3GPP interworking for positioning, the AMF will co-ordinate with the trusted non-3GPP gateway function (TNGF) or the non-3GPP interworking function (N3IWF), depending on whether the WiFi network is a trusted one or an open public (untrusted) one.

The enhanced LCS (eLCS) architecture supports three types of location requests. The mobile terminated location request (MT-LR) supports many of the commercial requests coming from an external client or a network function. Within this MT-LR, periodic, triggered or UE available types of requests are also supported. For example a deferred MT-LR request can be triggered by a UE moving into a pre-defined area. The network-induced location request (NI-LR) is used for emergency/law enforcement applications and the serving AMF initiates this request. The mobile originated location request (MO-LR) occurs when a UE requests the serving PLMN for its own location.

The QoS level of a localization request is also of importance in this discussion. Three types of localization QoS classes are supported in eLCS. In the Assured QoS class, if the LMF observes that the requested QoS level is not achieved, the result is discarded and an error message is sent toward the request originator. In the Best effort class, the LMF will send whatever result that is achieved to the request originator. A new QoS class, called the Multiple QoS class, was introduced in the Rel-17 eLCS work [13]. This class can have up to three accuracy levels (preferred, intermediate, and minimum) and if the location result meets any of these, LMF can pass the result toward the request originator. One of the many uses of this new class is when the request originator (or the client) is not sure of accuracy levels that will be achieved in intermediate stages of tracking a UE. For example, zonal localization can be used for tracking UEs over combined/linked temporal or spatial zones, where each zone can use a different localization method. The Multiple QoS class can ensure that a single localization request will yield valid results in any of these zones or at zone boundaries.

The privacy indication by the user (of the UE) is an essential requirement for any commercial localization request, as stipulated by eLCS architecture. When receiving a commercial localization request, the GMLC will check for the latest location privacy indication (LPI) in the stored UE information in UDM. If it is not present or up to date, the GMLC will terminate the request without proceeding. The LPI can be updated by the UE and this procedure is co-ordinated by the AMF and the results are passed onto the UDM.

3.2.2 Positioning Protocols

In the 5G context, positioning protocols are required to define formats for message exchanges between the target UE and the LMF. These messages can relate to UE or network element capability, any assistance data from the LMF and reporting of the positioning-related measurements or the actual position estimates [14]. In 5G, two types of positioning protocols are used, the LTE positioning protocol (LPP) and the new radio positioning protocol a (NRPPa). They are briefly explained below.

The LPP originated in the 4G (LTE) era and is now adopted and improved to support 5G positioning. Basically, LPP (in 5G) defines the message exchanges between the LMF and the UE (via the AMF) to initiate and conduct positioning. The LPP can be used for UE-assisted and UE-based positioning procedures. In UE-assisted

positioning, the UE reports the measurements and the LMF can use these to estimate the UE position. In UE-based positioning, the UE will estimate its position,
through the time- or angle-related reference signals provided by the network and
will report back the estimate to the LMF. The message flows for the initiation, provision of configuration data related to the reference signals (assistance data) by the
LMF, and reporting of the measurements or the location estimate by the UE are
all defined under the LPP. These specifications are contained in TS 37.355 [15].

The NRPPa is a new protocol defined in 5G and covers the message exchanges
between the NG-RAN and the LMF, via the AMF. The AMF connects to the
NG-RAN through the N2 reference point and the new NRPPa was developed
to carry the messages over the next-generation control plane interface (NG-C).
NRPPa is utilized in supporting network-assisted and network-based positioning
procedures. Both these procedures are executed by the serving gNB (or NG-eNB)
for the target UE. In network-assisted positioning, the serving gNB provides
only the measurements back to the LMF. Network-based positioning requires
the serving gNB to also estimate the location of the target UE and report this
back to the LMF. The NRPPa specifications are contained in TS 38.455 [16].
The configurations for the LPP and NRPPa protocols within the localization
architecture are shown in Figure 3.3.

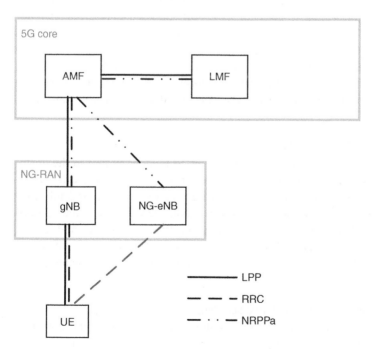

Figure 3.3 LPP and NRPPa configurations within 5G localization architecture [17].

3.2.3 RAT-Dependent NR Positioning Technologies

As both the network and the UE require each other's assistance for a proper estimation of UE's location, the standardization of the positioning methods is highly important. Some of these techniques have been already introduced and standardized in 4G, while then in 5G, they have been updated and enhanced with NR numerology. Other techniques have been introduced in the standard for the first time in 5G. These methods have enabled positioning advancements across many verticals and commercial use cases. In Rel-17, the target requirements are providing meter-level accuracies in public safety use cases or sub-meter accuracies in vehicle-to-everything (V2X) applications with 2D/3D coordinates [18]. These methods have also been developed to enhance the positioning requirements in industrial internet of things (IIoT) use cases to centimeter-level accuracies with end-to-end latency of less than 100 ms [19]. Within this section, we summarize the three main 3GPP 5G RAT-dependent categorization of positioning solutions.

3.2.3.1 Downlink-Based Solutions

The NR downlink-based positioning solutions are mainly referred to as positioning solutions in which the UE performs the positioning measurements based on some downlink signal from the gNB(s). These solutions can be timing-based techniques (i.e. downlink time-difference-of-arrival (DL-TDoA)), or angle-based techniques including downlink angle-of-departure (DL-AoD), or based on received reference signal power (PRS reference signal received power (PRS-RSRP)). There is also a downlink positioning reference signal (DL-PRS) defined in NR, which is further discussed in Section 3.2.4. Moreover, the CID, the carrier-phase-based techniques, and transmit/receive point (TRP), related information (e.g. reference signal (RS) resource and/or resource set ID) are also categorized within the downlink-based solutions.

3.2.3.2 Uplink-Based Solutions

NR uplink-based positioning solutions provide techniques similar to the downlink solutions listed above. In these methods, it is one or more gNB(s) which perform the positioning measurements based on uplink sounding reference signal (UL-SRS) sent by the UE. These solutions can be again timing-based techniques including uplink relative time-of-arrival (UL-RToA) and uplink time-difference-of-arrival (UL-TDoA), or angle-based techniques including uplink angle-of-arrival (UL-AoA), or based on received reference signal power (SRS reference signal received power (SRS-RSRP)).

3.2.3.3 Downlink- and Uplink-Based Solutions

The NR downlink- and uplink-based solutions use both uplink and downlink measurements to estimate the position of the target UE. One solution defined in NR is positioning based on multiple round trip time (multi-RTT) measurements or multiple antenna beam measurements to enable DL-AoD and UL-AoA estimates. While the multi-RTT positioning method is robust against network time synchronization errors, angle-based methods are more relevant with usage of mmWave and multiple antennas in 5G. For multi-RTT, the LMF initiates the procedure whereby multiple TRPs and a UE perform the gNB Rx-Tx and UE Rx-Tx measurements, respectively. For multi-RTT, gNBs and UEs transmit DL-PRS and UL-SRS, respectively. The RRC protocol is being used to configure UL-SRS to the UE, and the measurements from gNB are reported by NRPPa protocol to the LMF. This path is totally different for the downlink part, meaning that the LPP is used for both sending the DL-PRS to the UE and providing the measurements to the LMF.

3.2.4 Specific Positioning Signals

Multiple reference signals are used for communication-related procedures. For the purpose of RAT-based positioning methods, NR supports two new reference signals: the DL-PRS and the UL-SRS. Figure 3.4 shows an example within slot

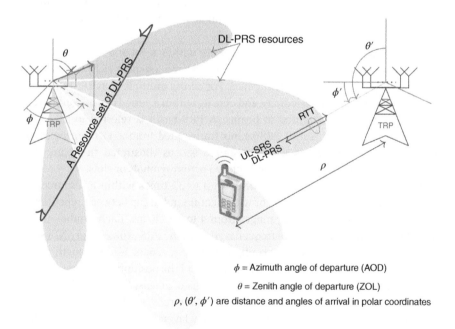

ϕ = Azimuth angle of departure (AOD)

θ = Zenith angle of departure (ZOL)

ρ, (θ', ϕ') are distance and angles of arrival in polar coordinates

Figure 3.4 An illustration of 5G positioning methods and elements such as beams as resources and sets of beams as resource sets [11].

configurations of DL-PRS and UL-SRS. The DL-PRS is used for the DL-TDoA method and the UL-SRS is used for the UL-TDoA method. Both signals are used for the multi-RTT and the angular measurements. The channel state information reference signal (CSI-RS) and single-sideband (SSB) signals used for radio resource management (RRM) can also be used as part of the eCID positioning method.

3.2.4.1 Downlink Positioning Reference Signal

The flexible bandwidth configuration allows the network to configure the DL-PRS while keeping out-of-band emissions to an acceptable level. The large bandwidth allows a very significant improvement in ToA accuracy compared to long term evolution (LTE). The beam structure of the PRS improves coverage, especially for mmWave deployments, and also allows for angle-of-departure (AoD) estimation; for example, the UE may measure DL-PRS RSTD per beam and report the measured RSTD including DL-PRS resource id (beam id) to the LMF.

The DL-PRS footprint on the time frequency grid is configurable with a starting physical resource block (PRB) and a PRS bandwidth. The PRS may start at any PRB in the system bandwidth and can be configured with a bandwidth ranging from 24 to 276 PRBs in steps of 4 PRBs. This amounts to a maximum bandwidth of about 100 MHz for 30 kHz sub carrier spacing to about 400 MHz for 120 kHz subcarrier spacing.

The DL-PRS can be configured at two levels: within a slot and at the multi slot level. For within a slot DL-PRS, the starting resource element (RE) in time and frequency from a TRP can be configured. For across multiple slots DL-PRS, gaps between PRS slots, their periodicity, and density within a period can be configured.

The PRS can be transmitted in beams. A PRS beam is referred to as a PRS resource, while the full set of PRS beams transmitted from a TRP on the same frequency is referred to as a PRS resource set, as illustrated in Figure 3.4. The different beams can be time-multiplexed across symbols or slots.

The DL-PRS resources can be repeated up to 32 times within a resource set period, either in consecutive slots or with a configurable gap between repetitions. The resource set period in FR1 ranges from 4 to 10,240 ms. Each symbol of the DL-PRS has a comb structure in frequency; that is, the PRS utilizes every Nth subcarrier. The comb value N can be configured to be 2, 4, 6, or 12. The length of the PRS within one slot is a multiple of N symbols, and the position of the first symbol within a slot is flexible as long as the slot consists of at least N PRS symbols.

The DL-PRS configuration is provided in a hierarchy, as shown in Figure 3.5. There can be at most four positioning frequency layers (PFL)s, and each PFL has at most 64 TRPs. Each TRP per frequency layer can have 2 DL-PRS resource sets, resulting in a total of eight resource sets per TRP, and each resource set can have

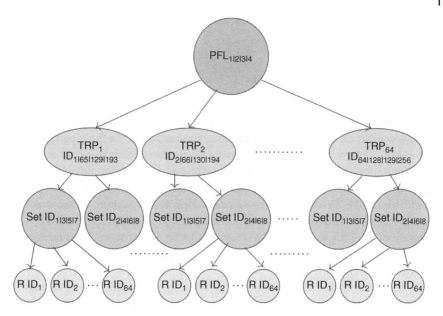

Figure 3.5 NR DL-PRS configuration hierarchy [11].

up to 64 resources. Each resource corresponds to a beam. Having two different resource sets per frequency layer per TRP allows gNB to configure one set of wide beams and another set of narrow beams for each frequency layer.

3.2.4.2 Uplink Signal for Positioning

The UL-SRS for positioning is a reference signal based on the SRS for communication. Although the signals have a lot in common, SRS for positioning and for communication are configured separately and with different properties specific to their usage.

The SRS for positioning is configured in a resource, which in turn can be part of a resource set. A resource corresponds to an SRS beam, and resource sets correspond to a collection of SRS resources (i.e. beams) aimed at a given TRP. The SRS resource is defined as a collection of symbols transmitted on the time–frequency NR grid.

Like the DL-PRS, the SRS resources for positioning are transmitted on a single antenna port, and can be placed to begin on any symbol in the NR uplink slot. In the time domain, the SRS resources for positioning can span 1, 2, 4, 8, 12 consecutive orthogonal frequency division multiplexing (OFDM) symbols, which provide enough coverage to reach all TRPs involved in the positioning procedures.

Contrary to the SRS for communication, repetition is not supported in an SRS for positioning resource. Similar to SRS for communication, the SRS for positioning uses Zadoff–Chu sequences as a base signal to ensure low-PAPR transmission from the UE.

3.2.5 Positioning Measurements

Accurate and timely measurements are necessary to ensure positioning; therefore, there are standardized requirements on the maximum time during which the measurements are to be performed and the maximum error for the reported measurements. The measurement period spans the time necessary to obtain no more than four measurement samples, while the achieved accuracy should not be worse than the corresponding measurement accuracy requirement, which is more stringent than in LTE. In order to improve positioning accuracy, more measurements can be collected. Measurements are collected per resource. Hence, repeated transmission of DL-PRS and UL-SRS resources helps to collect more measurements. The repetition of resources can be done in two ways, repeat before sweep and sweep before repeat.

In 5G/NR more downlink and uplink positioning measurements are standardized compared to 4G/LTE. In LTE, RSTD is the only measurement for PRS, while in NR, RSTD, UE Rx-Tx time difference and PRS-RSRP are downlink measurements based on DL-PRS that can be reported by the UE to the LMF. Similarly, UL-RToA, gNB Rx-Tx time difference, SRS-RSRP, azimuth angle-of-arrival (A-AoA), and zenith angle-of-arrival (Z-AoA) UL measurements can be configured by LMF and reported by a gNB. The UE may also be requested to report some RRM measurements such as synchronization-signal-based RSRP (SS-RSRP), SS-RSRQ, channel state information RSRP (CSI-RSRP), and CSI-RSRQ to support the eCID positioning method.

The 3GPP standard supports the aforementioned measurements in all supported bands and both lower and higher frequency ranges (FR1 and FR2, respectively) over the corresponding bandwidths within the operating frequency bands, which are up to 100 MHz in FR1 and 400 MHz in FR2. The range of reportable absolute values for power-based measurements is [−156; −31] dBm with 1 dB resolution, similar to that in LTE. The range of reportable absolute values for timing-based measurements is [−985,024; 985,024] in Tc units, with a flexible resolution step of 2 kTc, where 1 Tc corresponds to 0.51 ns, and k is an integer in the interval [2; 5] for FR1 and [0; 5] for FR2. The value of k can be configured by the LMF or adjusted by the UE. The above measurements can be performed for serving and neighbor TRPs, and can be used for a variety of RAT-dependent and hybrid positioning methods, standardized or not. Some examples of the standardized methods and corresponding measurements are provided in Table 3.2.

Table 3.2 5G standardized methods and corresponding measurements.

Positioning method	Positioning measurements
DL-TDoA	RSTD, optional complementary PRS-RSRP
DL-AoD	PRS-RSRP
UL-TDoA	UL-RToA, optional complementary SRS-RSRP
UL-AoA	A-AoA, Z-AoA, optional complementary SRS-RSRP
multi-RTT	gNB Rx-Tx and UE Rx-Tx time difference, optional complementary PRS-RSRP, SRS-RSRP, A-AoA, Z-AoA

3.3 Hybrid Positioning Technologies

Supporting hybrid positioning methods that integrate RAT-dependent and RAT-independent technologies will enable seamless and accurate positioning for several commercial use cases of the 5G ecosystem. Currently, there is no clear leading technology for hybrid positioning, and each technology has unique properties that can provide accurate localization in specific scenarios.

For trilateration to work correctly, a dense deployment is required for a single technology. Such a deployment may still fail at points where shadowing of too many reference points occurs, producing coverage holes, or at the border of the deployment. In such points, opportunistic range fusion can help better exploit the visible reference points. For instance, 5G can be used to complement the RTT systems, making location possible where only three or fewer reference points are visible even from different technologies. In what follows, two examples of fusion, one outdoors, using GNSS-RTK and 5G-NR, and one indoors, using 3GPP with UWB and WiFi-FTM, are described.

3.3.1 Outdoor Fusion

The combination of 5G positioning technologies and GNSS can fulfil high-accuracy positioning requirements in many future outdoor use cases [20]. While it is of interest to combine any of the 5G positioning features with GNSS and exploring the benefits of this hybrid positioning method, for a start, a simple

RTT of signal from the UE to the serving gNB in a 5G network is considered. The 5G RTT and GNSS pseudo-range errors in line-of-sight (LOS) conditions are assumed to be Gaussian-distributed with zero-mean and a certain error variance. These LOS measurements are then considered to assist a hybrid 5G RTT and multi-constellation GNSS solution in a deep urban canyon. The resulting horizontal positioning accuracy of the hybrid approach can significantly improve the GNSS stand-alone solutions, by using only one high-accuracy 5G RTT observable. The deep urban canyon conditions often limit the availability of LOS satellites even with multi-constellation solutions. Thus, the additional 5G RTT observable provides three main benefits to complement the GNSS solution, in terms of relaxation of the positioning problem, improved geometry, and enhanced observable.

3.3.2 Indoor Fusion

In the last few years, UWB has become a de-facto standard for indoor localization and has started to be commercialized in smartphones in early 2020. Thanks to its precision and robustness to interference and multipath, UWB can provide a location with an accuracy of less than 10 cm. But UWB is not the only clear alternative; WiFi-based localization, either with fingerprinting or with WiFi fine time measurement (FTM) in IEEE 802.11-2016, is a competing technology that is now available in commodity chipsets.[1] Fingerprinting exploits the ubiquity of WiFi APs in indoor scenarios, achieving high precision without the need to modify hardware deployments, but at the cost of requiring regular and manual offline training. WiFi FTM and UWB require the deployment of reference points. While UWB anchors are normally cheaper, WiFi FTM APs can offer connectivity along with localization. Both UWB and WiFi FTM have associated infrastructure costs, as at least four reference points must be visible at all times (three in the case of 2D localization), and their location must be known.

In indoor scenarios, it is common that multiple technologies coexist within the same area. For instance, wherever UWB is deployed, there is a high chance that some network connectivity is provided with WiFi, 5G, or both. In turn, this multiplies the number of potential reference points for localization. A hybrid scheme can be defined, where the different technologies are opportunistically used for estimating distances between the target and the reference points, the estimations are sent to a central localization service, and an estimation is done using trilateration methods such as least squares (LS).

---

1 Although WiFi location systems achieve remarkable accuracy, with decimeter-level errors for recent system designs, they require not only timing information, but also access to channel state information (CSI) to derive accurate angle, which cannot currently be leveraged by 5G RAT-independent methods.

3.4 5G Advanced Positioning

3GPP recognizes Rel-18 as the first release of 5G Advanced to highlight the significant evolution of the 5G system (5GS). One key component of 5G Advanced is the use of Artificial intelligence (AI) based on machine learning (ML) techniques. AI/ML is expected to trigger a paradigm shift in future wireless networks. AI/ML-based solutions will be used to introduce intelligent network management and solve multi-dimensional optimization issues with respect to real-time and non-real-time network operation. AI/ML will also be used to improve the radio interface by further optimizing the performance of complex multi-antenna systems, for example. New use cases such as extended reality (XR) communication will use wireless networks to provide immersive experiences in cyber–physical environments and enable human–machine interactions using wireless devices and wearables [21].

It is expected that AI/ML can significantly improve NR air-interface performance. The RAN standardization in Rel-18 explores the opportunities by setting up a general framework for AI/ML-related NR air-interface enhancements, including proper AI/ML modeling, evaluation methodologies, and performance requirements/testing. The first few areas for concrete AI/ML enhancements are beam management, channel estimation, and positioning.

Positioning integrity is a measure of the trust in the accuracy of the position-related data and the ability to provide timely warnings based on assistance data provided by the network. GNSS integrity was in the scope of Rel-17, and Rel-18 extends this to address other positioning techniques as well as other relevant integrity aspects of mission critical use cases that rely on positioning estimates and the corresponding uncertainty estimates. Integrity enables applications to make the correct decisions based on the reported position, e.g. when monitoring a robotic arm to decide whether its arm movement is within allowed limits to ensure safety distances to humans and other objects.

Rel-17 has specified support for reduced capability (RedCap) UEs with reduced bandwidth support and reduced complexity including reduced number of receive chains. Such UEs could support NR positioning functionality, but there is a gap in that the core and performance requirements have not been specified for the positioning-related measurements performed by RedCap UEs, and Rel-18 evaluates how the reduced capabilities of RedCap UEs might impact eventual position accuracy [22].

Moreover, the V2X and public safety use-cases requirements of in-coverage, partial-coverage, and out-of-coverage scenarios studied in Rel-17 and the requirements on the ranging-based services demanded 3GPP to study and develop sidelink positioning solution that can support the use cases, scenarios, and requirements identified during these activities.

References

1 X. Lin, J. Bergman, F. Gunnarsson, O. Liberg, S. M. Razavi, H. S. Razaghi, H. Rydn, and Y. Sui. Positioning for the Internet of Things: A 3GPP perspective. *IEEE Communications Magazine*, 55(12):179–185, 2017.

2 M. Z. Win and R. A. Scholtz. Ultra-wide bandwidth time-hopping spread-spectrum impulse radio for wireless multiple-access communications. *IEEE Transactions on Communications*, 48(4):679–691, 2000.

3 D. Dardari, A. Conti, J. Lien, and M. Z. Win. The effect of cooperation on localization systems using UWB experimental data. *EURASIP Journal on Advances in Signal Processing*, 2008:1–11, Article ID 513873, 2008. Special issue on *Cooperative Localization in Wireless Ad Hoc and Sensor Networks*.

4 A. Conti, M. Guerra, D. Dardari, N. Decarli, and M. Z. Win. Network experimentation for cooperative localization. *IEEE Journal on Selected Areas in Communications*, 30(2):467–475, 2012.

5 D. Dardari, A. Conti, U. J. Ferner, A. Giorgetti, and M. Z. Win. Ranging with ultrawide bandwidth signals in multipath environments. *Proceedings of the IEEE*, 97(2):404–426, 2009.

6 Decawave. DWM1001 System Overview and Performance, 2018. Accessed date: 29/08/2021 https://www.decawave.com/dwm1001/systemoverview/.

7 R. Mautz. Indoor positioning technologies. *ETH Zurich, Department of Civil, Environmental and Geomatic Engineering*, 2012.

8 E. T. S. I. (ETSI). Ultra Wide Band, 2020. Accessed date: 2/09/2021 https://www.etsi.org/technologies/ultra-wide-band?jjj=1596015945046.

9 C. Daily. Apple AirTags use UWB wireless tech. Here's how ultra wideband makes your life easier, 2020. Accessed date: 22/09/2021, https://www.cnet.com/tech/mobile/apple-airtags-use-uwb-wireless-tech-heres-how-ultra-wideband-makes-your-life-easier/.

10 A. Tahat, G. Kaddoum, S. Yousefi, S. Valaee, and F. Gagnon. A look at the recent wireless positioning techniques with a focus on algorithms for moving receivers. *IEEE Access*, 4:6652–6680, 2016.

11 S. Dwivedi, R. Shreevastav, F. Munier, J. Nygren, I. Siomina, Y. Lyazidi, D. Shrestha, G. Lindmark, P. Ernström, E. Stare, S. M. Razavi, S. Muruganathan, G. Masini, A. Busin, and F. Gunnarsson. Positioning in 5G networks. *IEEE Communications Magazine*, 59(11):38–44, 2021.

12 E. Dahlman and S. Parkvall. NR - The new 5G radio-access technology. In *2018 IEEE 87th Vehicular Technology Conference (VTC Spring)*, pages 1–6, 2018.

13 TS 23.273. 3rd Generation Partnership Project (3GPP), 5G system (5GS) location services (LCS); Stage 2 (Release 17), June 2021. Release 17.

14 TS 38.305. 3rd Generation Partnership Project (3GPP), Technical Specification Group Radio Access Network; NG Radio Access Network (NG-RAN); Stage 2 functional specification of user equipment (UE) positioning in NG-RAN, April 2022. Release 17.

15 TS 37.355. 3rd Generation Partnership Project (3GPP), LTE Positioning Protocol (LPP), 2022. Release 17.

16 TS 38.455. 3rd Generation Partnership Project (3GPP), NG-RAN; NR Positioning Protocol A (NRPPa), 2022. Release 17.

17 5G positioning: What you need to know, 2020. URL https://www.ericsson.com/en/blog/2020/12/5g-positioning–what-you-need-to-know.

18 TR 38.845. 3rd Generation Partnership Project (3GPP), Technical Report Group Radio Access Network; Study on scenarios and requirements of in-coverage, partial coverage, and out-of-coverage NR positioning use cases, October 2021. Release 17.

19 TR 38.857. 3rd Generation Partnership Project (3GPP), Technical Report Group Radio Access Network; Study on NR positioning enhancements, March 2021. Release 17.

20 J. A. del Peral-Rosado, F. Gunnarsson, S. Dwivedi, S. M. Razavi, O. Renaudin, J. A. López-Salcedo, and G. Seco-Granados. Exploitation of 3D city maps for hybrid 5G RTT and GNSS positioning simulations. In *ICASSP 2020 - 2020 IEEE International Conference on Acoustics, Speech and Signal Processing (ICASSP)*, pages 9205–9209, 2020.

21 Ericsson. Whitepaper, 5G advanced: Evolution towards 6G, June 2022.

22 TR 38.859. 3rd Generation Partnership Project (3GPP), Technical Report Group Radio Access Network; Study on expanded and improved NR positioning, January 2023. Release 18.

4

Enablers Toward 6G Positioning and Sensing

Joerg Widmer[1], Henk Wymeersch[2], Stefania Bartoletti[3], Hui Chen[2], Andrea Conti[4], Nicolò Decarli[5], Fan Jiang[2], Barbara M. Masini[5], Flavio Morselli[4], Gianluca Torsoli[4] and Moe Z. Win[6]

[1] *IMDEA Networks Institute, Madrid, Spain*
[2] *Department of Electrical Engineering, Chalmers University of Technology, Gothenburg, Sweden*
[3] *Department of Electronic Engineering and CNIT, University of Rome Tor Vergata, Rome, Italy*
[4] *Department of Engineering and CNIT, University of Ferrara, Ferrara, Italy*
[5] *National Research Council - Institute of Electronics, Computer and Telecommunication Engineering and WiLab-CNIT, Bologna, Italy*
[6] *Laboratory for Information and Decision Systems, Massachusetts Institute of Technology, Cambridge, MA, USA*

As 5G evolves into 5G Advanced and then into 6G, there are several technological advances that will enable more fine-grained positioning than in 5G, including the use of new spectrum, the tight integration of communication, positioning, and sensing, the introduction of reconfigurable intelligent surface (RIS), to name but a few. These advances will also provide high-resolution sensing information about passive objects, people, and the environment, which is a new service not present in 5G. The enablers are largely driven by the need for higher data rates and lower latencies, in support of new use case families. The only way to provide higher data rates is to consider new frequency bands, especially in the upper mmWave region (above 100–300 GHz, sometimes referred to as the lower THz band). At these bands, several GHz of bandwidth are available, though they come with significant challenges. First of all, the channel behaves differently than at lower frequencies, e.g. due to molecular absorption, and suffers from severe path loss. Secondly, radio frequency (RF) hardware is less well developed than at lower frequencies, making it difficult to generate sufficient output power efficiently. In addition, hardware impairments will affect communication and positioning in different ways. On the positive side, miniaturization at higher frequencies

Positioning and Location-based Analytics in 5G and Beyond, First Edition.
Edited by Stefania Bartoletti and Nicola Blefari Melazzi.
© 2024 The Institute of Electrical and Electronics Engineers, Inc. Published 2024 by John Wiley & Sons, Inc.

provides the opportunity to pack more antenna elements in a given footprint. The abundant bandwidth and antenna elements provide a combination of high delay and angle resolution, far beyond the capabilities of 5G.

In this chapter, we discuss three major trends toward 6G, namely *integrated sensing and communication (ISAC)*, the ability to control radio wave propagation in the environment thanks to *RISs*, and advances in *model-based and model-free signal processing*. This list of trends is by no means exhaustive, but only aims to highlight the tremendous potential for research and innovation toward 6G positioning.

4.1 Integrated Sensing and Communication

All communication systems (including 5G) have different sensing functionalities. Consequently, there are many ways in which sensing can be integrated with communication. In this section, we aim to provide an overview of these different perspectives. We can categorize sensing into several groups: (i) direct sensing for communications (which includes carrier sensing and channel estimation); (ii) sensing for positioning (which includes sending dedicated pilots and parametric channel estimation); (iii) radar-like sensing (which includes monostatic and bistatic configurations, device-free positioning, and synthetic aperture); (iv) non-radar-like sensing (which includes RF sensing and weather monitoring), often combined with machine learning methods. Similarly, we can categorize ISAC into different groups (loosely based on [1]):

- Simultaneous communication and monostatic sensing: This is arguably the most well-researched topic in ISAC, where a full-duplex transceiver sends a communication waveform to a remote receiver and at the same time processes the backscattered signal to obtain estimates of, e.g. range, angle, or Doppler, with respect to one or more targets [2]. Interestingly, it is also the case in which communication and sensing are in a direct trade-off [3], since the waveform can be optimized for communication (e.g. using waterfilling and directional transmission to the communication receiver) or sensing (e.g. using uniform power allocation and scanning beams), but not both at the same time.
- Sensing as a service: A prerequisite for simultaneous communication and monostatic sensing is that the same hardware and waveform are used for communication and sensing. However, this is not given for other types of sensing. 6G systems may prefer to have separate waveforms for sensing (e.g. frequency-modulated continuous wave [FMCW]) and for communication (e.g. orthogonal frequency division multiplexing [OFDM] or orthogonal time

frequency space [OTFS]) and schedule these functions at different times [4, 5]. In that case, sensing could be seen as a service that should not rely (only) on opportunistic transmissions; both rather provide a certain quality of service. Moreover, while joint hardware for communication and sensing is by now widely accepted, there may be scenarios in which dedicated sensing base stations may be deployed. All of these alternatives contribute to the richness (as well as confusion) surrounding ISAC in 6G.

- Sensing-aided communication: Once sensing information is available, it can be used to improve communication quality. One example is the detection of strong reflectors based on radar-like sensing, which could be harnessed for directional transmission if the LoS path to the user is interrupted [6]. Another example is the radar-aided tracking of users, so that once communication should be established, good prior knowledge is available, which can speed up initial access [7]. This is an extension on the more established research track from the 5G era of *location-aided communication* [8] (including location-based analytics), which involves using location information of connected users to improve communication quality. This can be at an instantaneous level, e.g. for beam tracking, at a medium-term level, e.g. predicting when a user will go into a tunnel and proactively downloading data, or at a long-term level, e.g. for base station deployment, avoiding drive tests by operators.

The remainder of this section describes two applications of ISAC. The first one (joint radar and communication with sidelink vehicle-to-everything [V2X]) falls in the category of *simultaneous communication and monostatic sensing* and highlights the potential to complement classical sensors with Beyond 5G sensing, provided the associated resource allocation challenges can be solved. The second application (human activity recognition and person identification) falls in the category of *sensing as a service* and highlights the potential to extract fine details regarding the activities and traits of individuals from high-resolution information within the received Beyond 5G signal. Table 4.1 lists the acronyms used in this chapter.

4.1.1 ISAC Application: Joint Radar and Communication with Sidelink V2X

As vehicles become connected, new opportunities but also new challenges are represented by the inclusion of sensing capabilities in vehicular communications, which can improve existing services (e.g. collective perception) and open the way to new services. Besides classical sensors such as cameras, lidars, and automotive radars, a further potentiality is offered by exploiting sidelink communication signals for radar sensing, thus enabling ISAC.

Table 4.1 List of acronyms

Acronym	Definition
3GPP	3rd Generation Partnership Project
AI	Artificial intelligence
AoA	Angle-of-arrival
AoD	Angle-of-departure
AP	Access point
BS	Base station
C-V2X	Cellular-V2X
CAM	Cooperative awareness message
CIR	Channel impulse response
CIS	Continuous intelligent surfaces
CNN	Convolutional neural network
CPM	Collective perception message
CRB	Cramér–Rao bound
CSI-RS	Channel state information reference signal
ECDF	Empirical cumulative density function
EM	Electro-magnetic
FMCW	Frequency-modulated continuous wave
ISAC	Integrated sensing and communication
LOS	Line-of-sight
MCS	Modulation and coding schemes
MIMO	Multiple-input-multiple-output
ML	Machine learning
MPC	Multi-path component
NLOS	Non-line-of-sight
OFDM	Orthogonal frequency division multiplexing
OTFS	Orthogonal time frequency space
PAPR	Peak-to-average power ratio
PRB	Physical resource block
QAM	Quadrature amplitude modulation
QPSK	Quadrature phase shift keying
RF	Radio frequency
RIS	Reconfigurable intelligent surface
RSRP	Reference signal received power
SCS	Subcarrier spacing

Table 4.1 (Continued)

Acronym	Definition
SI	Soft information
SLAM	Simultaneous localization and mapping
SNR	Signal-to-noise ratio
STFT	Short-time Fourier transform
UE	User equipment
V2X	Vehicle-to-everything
VRU	Vulnerable road user

4.1.1.1 V2X and Its Sensing Potential

Starting with Release 14 of 3rd Generation Partnership Project (3GPP), cellular-V2X (C-V2X) enables direct connectivity through the so-called sidelink mode, that is the direct communication among vehicles or between vehicles and other road actors (e.g. pedestrians, bikes, road side units) without going through the network infrastructure (i.e. via using the classical uplink or downlink). Sidelink communication is used to exchange several types of messages depending on the specific service. For example, each vehicle periodically broadcasts cooperative awareness messages (CAMs), which contain basic status information (such as identification, position, speed, acceleration, and direction). As another example, collective perception messages (CPMs) contain information of objects detected by onboard sensors and are used to enlarge the field of view of the vehicles. The frequency of generation and the size of the messages depend on the specific service (see Chapter 6).

The integration of sensing functionalities into vehicular communications requires challenging modifications, such as the use of full-duplex radios to allow simultaneous transmission and reception at the cost of employing mechanisms to avoid mutual interference. However, since vehicles are complex and expensive objects, the assumption of onboard embedded full-duplex radios with good self-interference cancellation is not unrealistic. In a scenario where a number of vehicles communicate with each other by relying on C-V2X radio access technology, each vehicle assumed to be equipped with a full-duplex radio, could be capable of listening to the backscattered response from the environment while transmitting messages (e.g. CAM or CPM) over V2X sidelink signals. By processing the response backscattered from the environment, each transmitter (transceiver) can detect the presence of targets in the surroundings (e.g. other non-cooperative vehicles, vulnerable road users (VRUs), obstacles) and estimate their distance and relative speed. This way, each vehicle performs radar operations using the V2X communication signal.

4.1.1.2 V2X Target Parameter Estimation and Signal Numerology

As we are assuming full-duplex radios, the radar receiver is co-located with the V2X transmitter and has perfect knowledge of the transmitted data symbols, which can be considered as pilots for the sake of parameter estimation. This way, the dependence of the received data from the transmitting data can be removed, by dividing each element of the received signal matrix by the transmitted signal matrix, as commonly assumed [9]. From the signal processing point of view, a simple yet effective approach for target detection is resorting to a double periodogram computed along the two directions corresponding to the sub-carriers and OFDM symbols. From the periodogram, coarse estimates of the target delay and Doppler shift can be obtained. Then, range estimation of the distance between the transmitting vehicle and the target is computed from the target delay; similarly, relative speed estimation of the relative target with respect to that of the transmitting vehicle is computed from the Doppler shift. The resolution of these coarse estimates is determined by the maximum bandwidth and overall signal duration. Therefore, the signal structure, modulation, and numerology impact the range and speed resolution. A short, non-exhaustive list of the main effects of modulation and numerology on sensing performance follows.

- Modulation: In 5G-V2X, several modulation formats can be adopted, such as quadrature phase shift keying (QPSK), 16 quadrature amplitude modulation (QAM), 64-QAM and 256-QAM with binary reflected Gray mapping, enabling increased spectral efficiency and throughput for high order modulations at the expense of sidelink coverage. The presence of random data symbols can affect the sensing performance, in particular when using non-constant modulus constellations, like high-order QAM, due to the increased noise floor for estimation of parameters. The actual effect on estimation depends on the processing implemented for removing the communication symbol dependence (e.g. division, reciprocal filtering, or matched filtering) [3].
- Subcarrier spacing and resource blocks: A transmission is composed of a slot in the time domain and a number of contiguous sub-channels in the frequency domain that depends on the size of the packet to transmit. The scalable numerology introduced by 5G-V2X implies flexible values for subcarrier spacing (SCS) and, increasing the SCS, the slot duration decreases, providing lower latency and higher resistance against Doppler effect and carrier frequency offset at high vehicular speeds and the bandwidth of a physical resource block (PRB) increases. Thus, the number of available resources in a given radio channel decreases, requiring higher modulation and coding schemes (MCSs)

to accommodate the message. Hence, the MCS impacts the number of PRBs needed for accommodating the packet (lower MCS index, higher number of PRBs) and, therefore, the transmission bandwidth. Increasing the MCS can thus have a detrimental effect on the sensing performance. In general, distance estimation performance is quite poor for distances larger than 100 m, due to the limited bandwidth considered today by regulation bodies (usually 10/20 MHz in the 5.9 GHz spectrum region).

- Medium access control: When multiple vehicles are present in the scenario and share the same wireless medium, performance is impacted by reciprocal interference. The medium access and resource allocation policies in the presence of multiple vehicles impact on the level of interference that each vehicle can experience (depending on vehicular density, transmission periodicity, packet size, resource allocation policy, etc.).

4.1.1.3 V2X Resource Allocation

Resource allocation mechanisms are critical when using V2X communications for ISAC. For example, in 5G-V2X, two sidelink modes are defined, named Modes 1 and 2, that correspond to controlled and autonomous operation modes, respectively. The autonomous mode is unquestionably the most challenging of the two modes as the channel sensing and distributed resource selection mechanisms play a critical role for an efficient and effective sharing of the sidelink resources among vehicles and therefore for the interference level. In Mode 2, the user equipment (UE) senses and decodes the sidelink control information sent by other UEs on the sidelink channel. Then, a decision on the set of resources to be used for the next packet transmission is taken.

An example scenario is illustrated in Figure 4.1. The decoded sidelink control information is stored together with the measurement of the reference signal

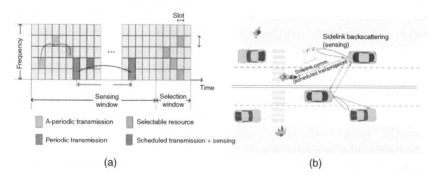

(a)   (b)

Figure 4.1 Resource allocation mechanism (a) and example scenario (b) for sidelink ISAC.

received power (RSRP), and this information is used to determine which resources must be excluded when a new resource selection is required. After the exclusion process, resources for transmission are selected randomly from the set of remaining available resources. When operating in the autonomous mode, it is likely that the same pool of resources can be shared (also in part) among different vehicles due to the distributed allocation mechanism. This fact produces interference that has an impact on the sensing performance of a vehicle aiming to use its sidelink signal for detecting the presence of targets in the surroundings.

4.1.2 ISAC Application: Human Activity Recognition and Person Identification

Future application scenarios for integrated sensing and communication will go beyond the mere tracking of objects. For example, several important use cases in the context of health care, smart factories, and smart homes involve human sensing, such as activity recognition and person identification.

4.1.2.1 Beyond Positioning

Such use cases require accurate fine-grained motion estimation, and for this reason, existing approaches for human sensing typically employ special-purpose, high-accuracy millimeter-wave radar devices [10]. However, with communication systems starting to use millimeter-wave frequencies and higher bandwidths, advanced ISAC designs are possible that can retrieve and analyze even the small-scale Doppler effect (micro-Doppler) caused by human motion with communication signals from base stations or access points [11]. To this end, information from the beam training procedure that aligns the directional antenna beams of millimeter-wave devices can be used to accurately localize and track individuals. Such beam training typically occurs on the order of hundreds of milliseconds, which is sufficient to track even highly mobile objects by observing changes in the channel impulse response (CIR). Once an object is located by observing changes in the reflected multi-path component (MPC) from it, efficient beam tracking or beam refinement mechanisms can be triggered at a smaller time scale than the beam training, to extract the desired micro-Doppler signatures from that specific MPC. The high spatial resolution provided by the large bandwidth of millimeter-wave communication systems allows even tracking and extracting the corresponding micro-Doppler signatures from multiple individuals simultaneously. This does not require any changes to the wireless network operation and entails negligible overhead to the communication rate, since beam training is required in any case by the communication devices, and additional beam refinement to measure the micro-Doppler is very efficient and requires very few resources compared to the actual data communication.

4.1.2.2 System Aspects

Such designs are compatible with any mmWave systems that support transmit beamforming for directional communication and CIR estimation. Such extraction of the CIR is feasible for the OFDM systems currently used for 5G-NR FR2, as well as IEEE 802.11ay WLANs at 60 GHz and future mobile systems at even higher frequencies that may use single carrier modulation schemes due to the simpler design and lower peak-to-average power ratio (PAPR) requirements. For the actual micro-Doppler extraction, 3GPP 5G-NR base stations can send frequent downlink channel state information reference signal (CSI-RS) to estimate the channel using different beam patterns, and IEEE 802.11ay systems provide in-packet training fields for CIR estimation for beam tracking purposes. For simplicity, we assume that the transmitter and the receiver units are co-located: the signal sent by the former, after bouncing off nearby reflectors (objects or humans), is collected at the receiver that retrieves information for each reflector, such as its distance and angle from the receiver, its velocity and micro-Doppler. Note that since CIR estimation is typically very robust and beam patterns are very directional, it may be suffi-cient to use one RF chain of a multiple-input-multiple-output (MIMO) base station as transmitter and another as receiver without any sophisticated self-interference cancellation, to avoid the complexity of a true full-duplex transceiver. The scheme can be extended to bi-static operation of sensing between multiple base stations.

4.1.2.3 Processing Chain (see Figure 4.2)

- Stage 1: Positioning and tracking: The CIR is key to inferring the positions of objects and persons in the environment. It is a vector of complex channel gains that provides information about the multi-path reflections of the signal, includ-ing their associated angle-of-arrival (AoA) and path delay. To infer the positions of persons it is key to remove the reflections from the static background, in particular since these often have a higher amplitude than those generated by humans. The background-related CIR is estimated by computing the time aver-age of the CIR amplitude over a window of static samples, as static reflections

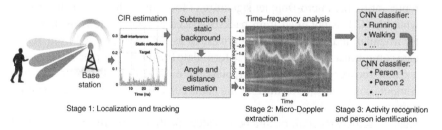

Stage 1: Localization and tracking   Stage 2: Micro-Doppler extraction   Stage 3: Activity recognition and person identification

Figure 4.2 Human activity recognition and person identification via micro-Doppler extraction.

are constant across time. Then, the CIR of the MPCs of interest is obtained by removing the amplitude of the static paths. We can then estimate the distance d_ℓ and AoA θ_ℓ of the corresponding reflector of path ℓ. The distance is simply given by $d_\ell = c\tau_\ell/2$, where c is the speed of light and τ_ℓ is the path delay. The CIR is estimated for each of the B beam patterns tested during beam training. We can then estimate the AoA θ_{ℓ_j} of MPC ℓ as

$$\theta_\ell = \arg\max_\theta \sum_{b=1}^B g_b(\theta) \frac{|h_{\ell,p}|^2}{\|\mathbf{s}_\ell\|_2}, \tag{4.1}$$

where $g_b(\theta) \in [0,1]$ is the normalized gain of beam pattern b along direction θ, $h_{\ell,b}$ is the CIR amplitude if path ℓ with beam pattern b, and the vector $\mathbf{s}_{\ell_j} = \left[|h_{\ell,1}|^2, |h_{\ell_j,1}|^2, \ldots, |h_{\ell,B}|^2\right]^T$, and $\|\cdot\|_2$ is the L_2-norm. The rationale behind (4.1) is that if $|\tilde{h}_{\ell_j,p}|$ originates from the signal reflected off a subject, the corresponding angular direction is the one leading to the highest correlation between the CIR amplitude and the set of beam pattern gains. Upon obtaining the distance and the angle estimates, a Kalman filter can be used to smoothly track the subject's positions over time. Further details of such a design can be found in [11].

- Stage 2: Micro-Doppler extraction: With the above, we can analyze the phase of the ℓth path over time to extract the micro-Doppler effect caused by the movement of the subject. Human movement causes a frequency modulation on the reflected signal due to the small-scale Doppler effect produced by the different body parts. Using time–frequency analysis of the received signal, it is possible to distinguish between different actions performed by a person or identify the individual based on his/her way of walking. The micro-Doppler effect of human movement can be extracted from subsequent estimates of the CIR of a path ℓ by computing the short-time Fourier transform (STFT) across the slow-time dimension, i.e. the CIRs collected across different packets. To capture the human movement evolution across time, we can compute the micro-Doppler vectors for a window of N subsequent time-steps and concatenate them into a spectrogram representing the micro-Doppler signature of the target. This is straightforward when the CIR is obtained from packets spaced at regular intervals, but even for irregular and sparse traffic patterns such as the ones obtained from actual data communication, it is possible to formulate the micro-Doppler extraction as a sparse recovery problem to interpolate the missing data [12]. Note that this computation is only performed for the paths corresponding to a reflection off a person.
- Stage 3: Activity recognition and person identification: Once the micro-Doppler signatures of each person have been separated, a convolutional neural network (CNN) can be trained to extract features from the micro-Doppler spectrograms

and to classify them by learning a function $F(\cdot)$ that maps a micro-Doppler window onto a vector containing the class probabilities for activities (e.g. walking, running, sitting, and waving hands). Thanks to the separation of the CIR of different individuals and the subsequent computation of the micro-Doppler per MPC, it is possible to recognize different activities performed simultaneously by multiple subjects within the same indoor space. Moreover, since the gait of a person has unique characteristic features, a second CNN module can be used to extract and analyze gait features from the micro-Doppler signature and thus identify a person among a set of individuals. Using only two access points (to deal with occlusion of one subject behind another), an activity recognition accuracy of 94% and person identification accuracy of 90% can be reached, as shown in experiments with up to five subjects [11].

4.2 Reconfigurable Intelligent Surfaces for Positioning and Sensing

Nearly passive RISs can be used to control electro-magnetic (EM) environment and thus serve as an important enabler for beyond 5G networks. They can be used to improve the accuracy of positioning and sensing systems due to their reconfigurability, low power consumption, and inexpensive fabrication [13]. Other recent findings on RIS-aided communications and positioning can be found in [14–17]. RISs are typically planar structures consisting of configurable small meta-atoms. Through the joint optimization of the RIS configurations and conventional transceiver designs, the system performance in terms of coverage, reliability, spectrum, and energy efficiency can be significantly improved. For positioning of UE and objects in the environment, RISs have been shown to play an important role in enabling and improving the desired system performance [18]. They can also be important for sensing applications, and thus form a natural complement for integrated sensing and communication. An overview of different uses of RIS for positioning and sensing is provided in Figure 4.3. The key idea of RIS-aided positioning and sensing is to carefully control the phase response of an RIS so that signals, scattered by different parts of the surface, superimpose coherently at the receivers. In particular, continuous intelligent surfaces (CISs) allow controlling the phase response function continuously over the entire surface. A near/far-field signal model and the theoretical limits for positioning and communications via continuous and discrete RISs are obtained in [18], together with optimal design of the phase response in different scenarios including those with obstacles and multiple surfaces. In a positioning and mapping system, three essential components are required: a reference system, measurements, and inference algorithms [19]. In general, RISs can be operated in transmitter/reflector/receiver

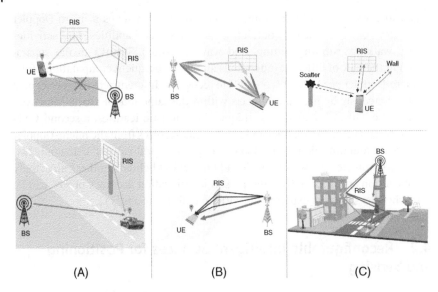

Figure 4.3 RIS enabling/enhancing positioning and sensing. (A) RIS enabling positioning, (B) RIS enhancing positioning, and (C) RIS for sensing.

modes, contributing to the positioning and mapping framework through the following aspects: (i) providing additional references if the RISs are deployed with known positions and orientations; (ii) offering additional location-dependent measurements due to the planar structures of the RISs; and (iii) developing novel inference algorithms by jointly optimizing the transceiver and RIS controllable unit cells [20].

4.2.1 RIS Enabling and Enhancing Positioning

The potential gain brought by the presence of RISs in a wireless environment, both in terms of positioning accuracy and system robustness, was determined in [16, 18]. In fact, the use of RISs can create a desirable electromagnetic environment for positioning in which the signals can be directed toward desired receivers and improve the quality of wireless measurements by controlling the phase response of RISs. Such control of the EM environment will create additional effective connections among agents and anchors in the network, improve the received signal-to-noise ratio (SNR), bridge indoor and outdoor environments, and enhance spatiotemporal cooperation. Compared with classical positioning, holographic positioning facilitates additional cooperation between RISs, agents, and anchors and leads to additional position-related information.

RISs can support existing positioning systems, an operation referred to as RIS-enhanced or RIS-boosted positioning. On the other hand, RISs can also provide positioning capabilities that would be impossible with the existing positioning system, an operation referred to as RIS-enabled positioning. We will now describe these in detail.

4.2.1.1 RIS Enabling Positioning

In 4G and 5G positioning, due to the precious synchronization challenges between the UE and the base station (BS), narrowband channels, and limited antenna aperture, multiple BSs or access points (APs) are required to localize the user devices. In beyond 5G, when RISs are deployed with known positions and orientations, additional references will be provided. Moreover, since the signal path from an RIS can be separated from the direct path from the BS (if it is present) and the two paths are affected by the same clock bias, each RIS is inherently synchronized to the BS. In addition, the planar structure of RISs can resolve the directions of the incoming and outgoing radio signals from/toward the RISs, providing the angle measurements for positioning. Moreover, since a large number of RIS elements can be optimally configured to forward the impinging wave, the signal strength of the non-line-of-sight (NLOS) path (from the RIS) can be augmented, improving the estimation quality of the NLOS components. As a result, the exploitation of RIS can help localize the UEs in the scenarios where the UEs cannot localize themselves [21].

In Figure 4.3, we show two RIS-enabled positioning examples:

- Positioning under blocked line-of-sight (LOS) with two RISs [22]: In the challenging LOS blockage case, a single antenna UE communicates with the single antenna BS through two RISs in a single carrier setup. The RISs are deployed on the walls with known positions, providing the references for UE positioning. By carefully designing the RIS phase profiles, the two NLOS paths can be well separated, each providing the angle-of-departure (AoD) to the UE. Therefore, the intersection of the two lines gives the UE position. In this setup, the synchronization of the UE and the BS is not required.
- Positioning with a single RIS and one BS [23]: Another enabling example is the positioning of a user in a single-input single-output system with a single planar RIS. Thanks to the deployment of the RIS, the configuration of the RIS phase profile can separate the LOS and the NLOS components in the received signal. With the estimations of the two delays and one 3D AoD pair (azimuth and elevation), the positioning of the vehicle and synchronization can be simultaneously achieved.

To summarize, the use of RIS in the positioning enabling scenarios can provide references and reliable NLOS position-dependent measurements.

4.2.1.2 RIS Enhancing Positioning

RISs usually consist of a large number of unit cells, which can implement the elementary functions of reflection, refraction, and focusing. These functions are useful in positioning-enhancing applications [24]. Due to the physical extent of the RIS, RISs will in many cases work in near-field [25]. The coherent processing of the received signal from the well-structured RIS elements can improve the positioning performance when we exploit the wavefront curvature. By adjusting the RIS phase profile, the RIS can work as a smart reflector providing additional position-dependent measurements, such as angles and delays. In addition, RISs can direct an impinging radio wave toward a specific location by acting as a transmission lens [20]. By doing that, the NLOS path signal from the RIS to UE can be energy-concentrated, improving the received SNR for delay and angle estimations. As a result of these combined effects, RISs have the potential to significantly improve the positioning performance.

In Figure 4.3B, we show two examples where the RISs can enhance the positioning performance.

- RIS as an additional AP [26]: Specifically, the estimation of the NLOS position-dependent measurements can be improved due to the use of RISs, leading to much improved positioning performance even in an asynchronous setup.
- RIS for near-field positioning [25]: In the second case, we highlight the application of the RISs in indoor positioning where the near-field effects are present. Since the array of the RIS elements is usually large compared to the propagation distance in the indoor applications. Also, the energy-focused transmissions of the positioning reference signal can reduce energy consumption, as well as EM radiation.

4.2.1.3 Use Cases
The advantages of RISs have been quantified in terms of the positioning error bound of wireless sensing and spectral efficiency of communication. As example results, Figure 4.4 shows that the positioning accuracy with a RIS is superior to that without RIS, and that such performance gain increases with the carrier frequency. Figure 4.5 shows the significant performance gain in the spectral efficiency of the system with the aid of RIS compared to that without the aid of RIS, especially when the carrier frequency is high. From these example results, it is possible to expect that in future mobility systems supporting mm-Wave/THz bands and high bandwidth, RIS can provide a significant improvement in both positioning accuracy and spectral efficiency.

4.2.2 RIS for Sensing

In addition to improving the positioning performance of the UE, RISs can be deployed to improve the sensing performance [27], in terms of the detection and positioning of the stationary and moving objects in the environment. For example,

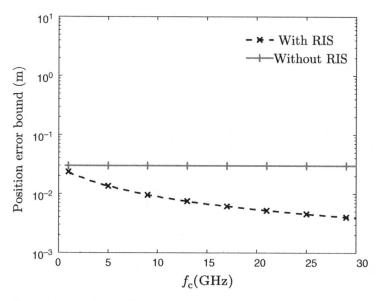

Figure 4.4 Benefits of RIS on positioning accuracy as a function of carrier frequency f_c.

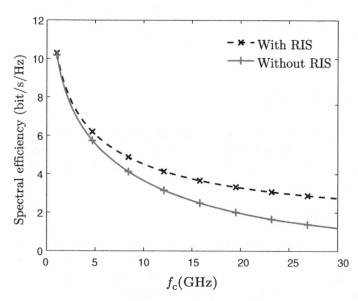

Figure 4.5 Benefits of RIS on spectral efficiency as a function of carrier frequency f_c.

in the monostatic radar scenario, the walls/windows can be coated with RISs. By designing the phase profile of the RISs, these walls/windows can be uniquely identified from the received signal. In addition, since the RIS can intelligently focus on the outgoing electromagnetic wave, the detection of the reflecting echos at the receiver side will be more accurate compared to the scattering case without RISs. In the bistatic radar case, RISs help the detection of NLOS path reflected/scattered from the objects due to the improved receive SNR. As a result, the positioning and mapping performance can be significantly improved, boosting the sensing applications [19]. In Figure 4.3C, we showcase two examples of RIS for sensing:

- UE as a personal radar with a single RIS: In this case, the deployed RIS acts as a reference, and no base station is required for localizing the UE [28]. The configuration of the RIS phase profile can resolve the reflected echo from RIS, while the echos from other objects are used to identify and map the objects in the environment.
- RISs for autonomous driving in urban cities: The blockage of the LOS path in urban cities poses challenges of radio simultaneous localization and mapping (SLAM) for autonomous driving. Meanwhile, the radio SLAM and communications are important functions for autonomous driving applications. By introducing the RISs, which can be deployed on the facades of large buildings, smart reflectors will be formed to forward the signal from/to the roadside units. The mobile vehicle can localize itself and map the objects in the environment based on the receive MPCs [29].

4.3 Advanced Methods

In addition to integrated sensing and communication and the use of reconfigurable intelligent surfaces, 6G will also feature novel ways to generate and process signals. The corresponding methods can be broken down into model-based approaches and AI-based approaches.

4.3.1 Model-Based Methods

Model-based methods rely on explicit models of the waveforms, channels, and system geometry to design and optimize systems for positioning, and process signals for minimizing positioning errors. Below, several examples of advances in model-based methods are presented, visualized in Figure 4.6.

- User-specific waveform optimization: Signal design, or waveform optimization, is an important component that affects positioning performance. The optimization of the waveform can be done in the time domain, frequency domain,

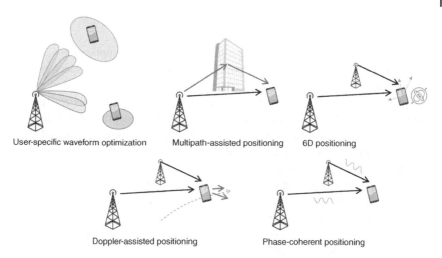

Figure 4.6 Five examples for advances in model-based methods toward 6G positioning.

and spatial domain. In the 5G NR specifications [30], the positioning reference signals have been introduced in the time–frequency domain. With prior UE position information, the spatial domain can also be exploited by using convex optimization tools to optimize the precoders at the BS to achieve a lower position error bound [31].

• Multipath-assisted positioning: While RISs provide controlled multipath, which has clear benefits for positioning and sensing, also the passive, uncontrolled multipath can be harnessed. Positioning using high-frequency systems (e.g. 5G and beyond) has the advantage of high temporal and spatial resolutions due to large bandwidth and antenna array size, which makes NLOS paths resolvable [32]. By estimating the channel geometrical parameters formulated in Chapter 3, these resolved NLOS paths can assist in mapping the environment (e.g. the incidence points) and estimating the position (and possibly orientation if equipped with an antenna array [33]) of the UE even with a single non-synchronized BS, which largely reduces the complexity of multiple BSs deployments. However, the benefit from NLOS paths depends also on the environment (e.g. number of paths, and the material of the reflection surfaces) [34].

• 6D Positioning: Most positioning works are limited to 2D or 3D space by assuming the UE is equipped with a single antenna. This is true in low-frequency band systems where the path propagation loss is low, and an isotropic antenna provides coverage to receive the signals from different directions. However, when moving to the higher frequency systems, an antenna array could also be available on the UE side, enabling orientation estimation at the UE. Useful tools such as constrained Cramér–Rao bound (CRB) and manifold optimization can be used for 3D orientation performance analysis and estimation [35].

- Doppler-assisted positioning: Positioning under mobility applications can be found in many scenarios, such as autonomous driving and high-speed trains, which degrades the positioning accuracy when done without considering the Doppler effect. Without increasing the complexity of the channel model too much, the Doppler shift estimation can provide additional direction information, which is determined by the velocity and the time duration of the signal [36]. Note that the coherence time and near-field model should not be ignored in Doppler-assisted positioning.

- Phase-coherent processing: In general, delay estimation is usually used for positioning and phase information of the channel gain is not utilized due to the environmental change and hardware imperfection. However, when phase-coherence can be maintained and phase cycle ambiguity can be resolved, a much better distance accuracy can be achieved [37]. The relevant standard mechanisms and algorithms to use the carrier phase for high-precision positioning are defined in 5G-advanced NR networks, and the numerical results show significant positioning accuracy improvement [38].

4.3.2 AI-Based Methods

The 6G wireless networks will integrate ML and artificial intelligence (AI)-based approaches by design to improve the performance of both communication and positioning. In particular, the use of ML techniques allows learning directly from the environment where the network will operate, instead of relying on analytically derived models which may not accurately describe the phenomena of interest. This is especially important considering that 6th generation (6G) networks will exploit THz bands, where current models fail to properly account for all the effects introduced by the use of such extremely high frequencies.

A variety of ML methods can be used to improve positioning, including supervised ML, unsupervised ML, and reinforcement learning methods. As an example, SI-based positioning briefly described in Chapter 2 leverages unsupervised ML techniques and shows numerous advantages for positioning in 6G networks. SI-based approaches aim at learning complex statistical model characterizing the relationship between the network measurements and the device position. This approach can efficiently fuse heterogeneous information (both from different types of measurements and contexts) and can be implemented in a distributed manner. Both aspects are particularly important for 6G networks, where a large amount of heterogeneous data can be used to infer the device positions and multiple devices can collaborate to perform distributed tasks. In addition to the aforementioned characteristics, ML-based positioning typically outperforms classical algorithms in terms of positioning accuracy.

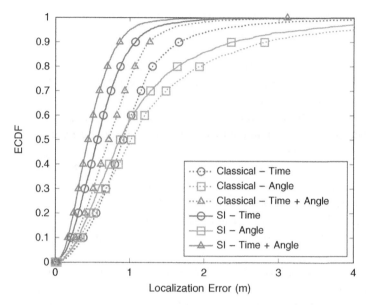

Figure 4.7 Improved positioning accuracy provided by soft information (SI)-based approaches (machine learning (ML)-based approaches) compared to classical ones considering the fusion of heterogeneous measurements.

4.3.2.1 Use Case

As example results, Figure 4.7 compares positioning performance for classical and ML-based approaches when considering the fusion of heterogeneous measurements, i.e. time- and angle-based measurements. In particular, comparison of a least square-like approach and SI-based approach is presented based on simulation of a 6G relevant scenario, i.e. millimeter-wave in an indoor environment. Performance is reported in terms of empirical cumulative density function (ECDF) of the positioning error. It can be observed that ML-based approaches significantly outperform the classical approaches considering both positioning based on a single measurement or the fusion of both and provides sub-meter accuracy at the 90th percentile.

References

1 F. Liu, Y. Cui, C. Masouros, J. Xu, T. X. Han, Y. C. Eldar, and S. Buzzi. Integrated sensing and communications: Towards dual-functional wireless networks for 6G and beyond. *IEEE Journal on Selected Areas in Communications*, 40(6):1728–1767, 2022.

2 F. Liu, C. Masouros, A. P. Petropulu, H. Griffiths, and L. Hanzo. Joint radar and communication design: Applications, state-of-the-art, and the road ahead. *IEEE Transactions on Communications*, 68(6):3834–3862, 2020.

3 M. F. Keskin, V. Koivunen, and H. Wymeersch. Limited feedforward waveform design for OFDM dual-functional radar-communications. *IEEE Transactions on Signal Processing*, 69:2955–2970, 2021.

4 L. Han and K. Wu. Joint wireless communication and radar sensing systems–state of the art and future prospects. *IET Microwaves, Antennas and Propagation*, 7(11):876–885, 2013.

5 P. Kumari, J. Choi, N. González-Prelcic, and R. W. Heath. IEEE 802.11ad-based radar: An approach to joint vehicular communication-radar system. *IEEE Transactions on Vehicular Technology*, 67(4):3012–3027, 2017.

6 A. Ali, N. Gonzalez-Prelcic, R. W. Heath, and A. Ghosh. Leveraging sensing at the infrastructure for mmWave communication. *IEEE Communications Magazine*, 58(7):84–89, 2020.

7 W. Yuan, F. Liu, C. Masouros, J. Yuan, D. W. K. Ng, and N. González-Prelcic. Bayesian predictive beamforming for vehicular networks: A low-overhead joint radar-communication approach. *IEEE Transactions on Wireless Communications*, 20(3):1442–1456, 2020.

8 R. Di Taranto, S. Muppirisetty, R. Raulefs, D. Slock, T. Svensson, and H. Wymeersch. Location-aware communications for 5G networks: How location information can improve scalability, latency, and robustness of 5G. *IEEE Signal Processing Magazine*, 31(6):102–112, 2014.

9 C. Sturm and W. Wiesbeck. Waveform design and signal processing aspects for fusion of wireless communications and radar sensing. *Proceedings of the IEEE*, 99(7):1236–1259, 2011.

10 J. Pegoraro, F. Meneghello, and M. Rossi. Multiperson continuous tracking and identification from mm-Wave Micro-Doppler signatures. *IEEE Transactions on Geoscience and Remote Sensing*, 59(4):2994–3009, 2020.

11 J. Pegoraro, J. O. Lacruz, E. Bashirov, M. Rossi, and J. Widmer. RAPID: Retrofitting IEEE 802.11ay access points for indoor human detection and sensing. *IEEE Transactions on Mobile Computing*, 2023.

12 J. Pegoraro, J. O. Lacruz, M. Rossi, and J. Widmer. SPARCS: A sparse recovery approach for integrated communication and human sensing in mmWave systems. In *ACM/IEEE International Conference on Information Processing in Sensor Networks (IPSN)*, pages 79–91, 2022.

13 M. D. Renzo, A. Zappone, M. Debbah, M.-S. Alouini, C. Yuen, J. de Rosny, and S. Tretyakov. Smart radio environments empowered by reconfigurable intelligent surfaces: How it works, state of research, and road ahead. *IEEE Journal on Selected Areas in Communications*, 38(11):2450–2525, 2020.

14 A. Elzanaty, A. Guerra, F. Guidi, and M.-S. Alouini. Reconfigurable intelligent surfaces for localization: Position and orientation error bounds. *IEEE Transactions on Signal Processing*, 69:5386–5402, 2021.

15 Y. Han, W. Tang, S. Jin, C.-K. Wen, and X. Ma. Large intelligent surface-assisted wireless communication exploiting statistical CSI. *IEEE Transactions on Vehicular Technology*, 68(8):8238–8242, 2019.

16 M. Z. Win, Z. Wang, Z. Liu, Y. Shen, and A. Conti. Location awareness via intelligent surfaces: A path toward holographic NLN. *IEEE Vehicular Technology Magazine*, 17(2):37–45, 2022 https://doi.org/10.1109/MVT.2022 .3157067. Special issue on *Backscatter and Reconfigurable Intelligent Surface Empowered Wireless Communications in 6G*.

17 Q. Wu and R. Zhang. Intelligent reflecting surface enhanced wireless network via joint active and passive beamforming. *IEEE Transactions on Wireless Communications*, 18(11):5394–5409, 2019.

18 Z. Wang, Z. Liu, Y. Shen, A. Conti, and M. Z. Win. Location awareness in beyond 5G networks via reconfigurable intelligent surfaces. *IEEE Journal on Selected Areas in Communications*, 40(7):2011–2025, 2022 https://doi. org/10.1109/JSAC.2022.3155542. Special issue on *Integrated Sensing and Communication*.

19 H. Wymeersch, J. He, B. Denis, A. Clemente, and M. Juntti. Radio localization and mapping with reconfigurable intelligent surfaces: Challenges, opportunities, and research directions. *IEEE Vehicular Technology Magazine*, 15(4):52–61, 2020 https://doi.org/10.1109/MVT.2020.3023682.

20 M. Di Renzo, A. Zappone, M. Debbah, M.-S. Alouini, C. Yuen, J. de Rosny, and S. Tretyakov. Smart radio environments empowered by reconfigurable intelligent surfaces: How it works, state of research, and the road ahead. *IEEE Journal on Selected Areas in Communications*, 38(11):2450–2525, 2020 https://doi.org/10.1109/JSAC.2020.3007211.

21 K. Keykhosravi, B. Denis, G. C. Alexandropoulos, Z. S. He, A. Albanese, V. Sciancalepore, and H. Wymeersch. Leveraging RIS-enabled smart signal propagation for solving infeasible localization problems. *IEEE Vehicular Technology Magazine*, 18(2):20–28, 2023.

22 E. Björnson, H. Wymeersch, B. Matthiesen, P. Popovski, L. Sanguinetti, and E. de Carvalho. Reconfigurable intelligent surfaces: A signal processing perspective with wireless applications. *IEEE Signal Processing Magazine*, 39(2):135–158, 2022 https://doi.org/10.1109/MSP.2021.3130549.

23 K. Keykhosravi, M. F. Keskin, G. Seco-Granados, P. Popovski, and H. Wymeersch. RIS-enabled SISO localization under user mobility and spatial-wideband effects. *IEEE Journal of Selected Topics in Signal Processing*, 16(5):1125–1140, 2022 https://doi.org/10.1109/JSTSP.2022.3175036.

24 J. He, F. Jiang, K. Keykhosravi, J. Kokkoniemi, H. Wymeersch, and M. Juntti. Beyond 5G RIS mmWave systems: Where communication and localization meet. *IEEE Access*, 10:68075–68084, 2022 https://doi.org/10.1109/ACCESS.2022 .3186510.

25 Z. Abu-Shaban, K. Keykhosravi, M. F. Keskin, G. C. Alexandropoulos, G. Seco-Granados, and H. Wymeersch. Near-field localization with a reconfigurable intelligent surface acting as lens. In *ICC 2021 - IEEE International Conference on Communications*, pages 1–6, 2021. doi: https://doi.org/10.1109/ ICC42927.2021.9500663.

26 A. Fascista, M. F. Keskin, A. Coluccia, H. Wymeersch, and G. Seco-Granados. RIS-aided joint localization and synchronization with a single-antenna receiver: Beamforming design and low-complexity estimation. *IEEE Journal of Selected Topics in Signal Processing*, 16(5):1141–1156, 2022 https://doi.org/10 .1109/JSTSP.2022.3177925.

27 X. Wang, Z. Fei, J. Huang, and H. Yu. Joint waveform and discrete phase shift design for RIS-assisted integrated sensing and communication system under Cramér-Rao bound constraint. *IEEE Transactions on Vehicular Technology*, 71(1):1004–1009, 2021.

28 K. Keykhosravi, G. Seco-Granados, G. C. Alexandropoulos, and H. Wymeersch. RIS-enabled self-localization: Leveraging controllable reflections with zero access points. *ICC 2022 – IEEE International Conference on Communications*, Seoul, Korea, Republic of, 2022, pp. 2852–2857.

29 H. Zhang, B. Di, K. Bian, Z. Han, H. V. Poor, and L. Song. Toward ubiquitous sensing and localization with reconfigurable intelligent surfaces. *Proceedings of the IEEE*, 2022.

30 TS 38.211. Technical Specification group radio access network; NR; physical channels and modulation (Release 17), October 2022. Release 17.

31 M. F. Keskin, F. Jiang, F. Munier, G. Seco-Granados, and H. Wymeersch. Optimal spatial signal design for mmWave positioning under imperfect synchronization. *IEEE Transactions on Vehicular Technology*, 71(5):5558–5563, 2022.

32 R. Mendrzik, H. Wymeersch, G. Bauch, and Z. Abu-Shaban. Harnessing NLOS components for position and orientation estimation in 5G millimeter wave MIMO. *IEEE Transactions on Wireless Communications*, 18(1):93–107, 2018.

33 M. A. Nazari, G. Seco-Granados, P. Johannisson, and H. Wymeersch. MmWave 6D radio localization with a snapshot observation from a single BS. *IEEE Transactions on Vehicular Technology*, 72(7):8914–8928, 2023.

34 F. Wen, J. Kulmer, K. Witrisal, and H. Wymeersch. 5G positioning and mapping with diffuse multipath. *IEEE Transactions on Wireless Communications*, 20(2):1164–1174, 2020.

35 M. A. Nazari, G. Seco-Granados, P. Johannisson, and H. Wymeersch. 3D orientation estimation with multiple 5G mmWave base stations. In *ICC 2021-IEEE International Conference on Communications*, pages 1–6. IEEE, 2021.

36 Y. Han, Y. Shen, X.-P. Zhang, M. Z. Win, and H. Meng. Performance limits and geometric properties of array localization. *IEEE Transactions on Information Theory*, 62(2):1054–1075, 2015.

37 H. Dun, C. C. Tiberius, and G. J. Janssen. Positioning in a multipath channel using OFDM signals with carrier phase tracking. *IEEE Access*, 8:13011–13028, 2020.

38 A. Fouda, R. Keating, and H.-S. Cha. Toward cm-level accuracy: Carrier phase positioning for IIoT in 5G-advanced NR networks. 2022 IEEE 33rd Annual International Symposium on Personal, Indoor and Mobile Radio Communications (PIMRC), Kyoto, Japan, 2022, pp. 782–787.

5

Security, Integrity, and Privacy Aspects

Stefania Bartoletti[1], Giuseppe Bianchi[1], Nicola Blefari Melazzi[1],
Domenico Garlisi[2], Danilo Orlando[3], Ivan Palamá[1] and Sara Modarres
Razavi[4]

[1] *Department of Electronic Engineering and CNIT, University of Rome Tor Vergata, Rome, Italy*
[2] *Department of Mathematics and Computer Science and CNIT, University of Palermo, Palermo, Italy*
[3] *University "Niccolò Cusano", Rome, Italy*
[4] *Ericsson Research, Ericsson AB, Stockholm, Sweden*

In the incoming fifth generation mobile communication network (5G), localization services, which in the past were mainly provided by non-cellular technologies, are now combined with solutions based upon cellular technologies and integrated within 5G architecture [1]. As a consequence, 5G architecture can improve the performance of the localization system but at the cost of new security and privacy vulnerabilities that need to be investigated and mitigated. This chapter provides a robust and secure approach for data sharing and provision by users and services. It identifies security and privacy challenges that apply to objects, systems, and networks and determine how to best address these challenges in 5G use cases and services.

This chapter addresses the research activities for what concerns the security and privacy aspects to integrate in new 5G ecosystem. Specifically, this chapter considers privacy mitigation techniques in new Location-Based Services (LBS), and we present a study of the robustness of different commercial smartphone related to the detection of the presence or the arrival of a user in a certain area, as well as the user's identification. Additionally, the chapter develops functions and algorithms to detect attacks over the air interface providing alerts that can be used for mitigation purposes. Finally, we also refer to the 3rd Generation Partnership Project (3GPP) positioning integrity aspects.

Positioning and Location-based Analytics in 5G and Beyond, First Edition.
Edited by Stefania Bartoletti and Nicola Blefari Melazzi.
© 2024 The Institute of Electrical and Electronics Engineers, Inc. Published 2024 by John Wiley & Sons, Inc.

5.1 Location Privacy

This section provides an overview on the privacy implication for cellular networks, and the functions that can be selected to include the privacy concerns into the complete lifecycle of the cellular architecture. Privacy threat happens when an attacker can create an association between identity, request content, and location of a user [2]. This information can be possibly obtained from attack to the network, specifically when the background knowledge is available.

In some cases, individuals may be unaware of the potential risks implied by the use of these services and of who is being permitted access to their location information. It is also known that telecommunication operators use location/position of their users both to optimize operations and to drive operational business opportunities [3]. For example, in 2014, Verizon built data centers in California to implement precision marketing with location data [4]. Thus, privacy has become a primary concern for LBSs.

The goal of the proposed approaches is to define techniques for secure authentication, data anonymization, and privacy preservation, so that all relevant actors can deal with private and secure data without compromising confidentiality, identities, and privacy of the users. In this section, we first present the International Mobile Subscriber Identity (IMSI) catcher attack, whose goal is to collect the identification code of the user terminal. We report a result evaluation that evaluates how different user terminals are robust with respect to the IMSI catcher attack, and we present a preliminary study to improve 5G authentication issues. Afterward, we provide an overview on the privacy implication for LBSs in cellular network, and thus, we propose possible solutions that can be applied to reduce system vulnerability. Table 5.1 lists the acronyms used in this Chapter.

5.1.1 Overview on the Privacy Implication

There is a privacy threat whenever an adversary can associate the identity with a user to information that the user considers private, including its location position [2]. Indeed, when location technology is significantly improved in 5G, it turns out that the risk of privacy violation will be increased too. Additionally, several new devices and gadgets will have positioning and tracking capabilities (location-based services) [5, 6]. To face this issue, the 3GPP has identified the following essential requirements related to user privacy [7]:

1. User identity confidentiality: The permanent identity of a user to whom a service is delivered cannot be eavesdropped on the radio access link.
2. User location confidentiality: The presence or the arrival of a user in a certain area cannot be determined by eavesdropping on the radio access link.
3. User untraceability: An intruder cannot deduce whether different services are delivered to the same user by eavesdropping on the radio access link.

Table 5.1 List of acronyms.

Acronym	Definition
3GPP	3rd Generation Partnership Project
5G	Fifth-generation
AKA	Authentication and key agreement
AL	Alert limit
AN	Access network
AoA	Angle-of-arrival
CRB	Cramér–Rao bound
DOS	Denial of service
ECIES	Elliptic-curve-integrated encryption scheme
GNSS	Global navigation satellite system
GSM	Global system for mobile communications
GUTI	Global unique temporary identifier
IMSI	International mobile subscriber identity
KPI	Key performance indicator
LBS	Location-based service
LMF	Location management function
MBR	Minimum bounding rectangle
MCC	Mobile Country code
MLE	Maximum likelihood estimation
MNC	Mobile network code
MSE	Minimum squared error
MSIN	Mobile subscriber identification number
NLJ	Noise-like jammer
NSA	Non-standalone
OS	Operating system
PL	Protection level
RAT	Radio access technology
RSS	Received signal strength
RSSI	Received signal strength indicator
SA	Standalone
SIDF	Subscription identifier deconcealing function
SIM	Subscriber identity module

(Continued)

Table 5.1 (Continued)

Acronym	Definition
SNR	Signal-to-noise ratio
SPEB	Square position error bound
SUCI	Subscription concealed identifier
SUPI	Subscription permanent identifier
SDR	Software-defined radio
TAU	Tracking area update
TBS	Terrestrial beacon system
TDoA	Time-difference-of-arrival
TIR	Target integrity risk
TMSI	Temporary mobile subscriber identity
ToA	Time-of-arrival
TTA	Time to alert
UE	User equipment
USRP	Universal software radio peripheral
WLAN	Wireless local area network

5.1.2 Identification and Authentication in Cellular Networks

Authentication is one of the building blocks of privacy measures in cellular networks. Authentication process has been a primary requirement in cellular networks since the old GSM (2G), and 5G has further invested in it by standardizing a novel public key-based approach to conceal the subscriber identity.

In mobile telephony systems, networks allocate to each Subscriber Identity Module (SIM) card a unique identifier, referred to as International Mobile Subscriber Identity (IMSI) in 4G and Subscription Permanent Identifier (SUPI) in 5G. As the authentication between a user and its service provider is based on a shared symmetric key, it can only take place after user identification. The IMSI is a unique number residing in a SIM card. An IMSI as defined in [8] is usually a string of 15 decimal digits. The first three digits represent the mobile Country code (MCC), while the next two or three form the Mobile Network Code (MNC) identifying the network operator. The remaining (nine or ten) digits are known as Mobile Subscriber Identification Number (MSIN) and represent the individual user of that particular operator. The 3GPP defines an Authentication and Key Agreement (AKA) protocol as well as procedures that support entity authentication, message integrity, and message confidentiality in addition to other security properties [9].

When the IMSI/SUPI values are sent in plain text over the radio access link, then users can be identified, located, and tracked using these permanent identifiers. To avoid this privacy breach, temporary identifiers are assigned to the SIM card by the visited network. These identifiers are called Temporary Mobile Subscriber Identity (TMSI) in 3G systems and Globally Unique Temporary User Equipment Identity (GUTI) in both 4G and 5G systems. Moreover, there exist certain situations where authentication using temporary identifiers is not possible. For instance in 4G, when a user registers with a network for the first time and is not yet assigned a temporary identifier. Another case occurs when the visited network is unable to resolve the IMSI/SUPI from the presented TMSI/GUTI.

An active man-in-the-middle adversary can intentionally simulate this scenario to force an unsuspecting user to reveal its long-term identity. These attacks are known as IMSI catching attacks [10] and persist in today mobile networks including the 4G LTE/LTE+ [11].

Additionally, although a temporary identifier may be used to hide a subscriber's long-term identity, researchers have shown that GUTI allocation is flawed: GUTIs are not changed as frequently as necessary [12], and GUTI allocation is predictable (e.g. with fixed bytes) [13].

5.1.3 IMSI Catching Attack

IMSI catchers can be used for (i) mass-surveillance of individuals in a geographical area, (ii) linking a real person to their identity in the network, (iii) tracing a person with a known IMSI, or (iv) checking their presence in a building or area. The attack is based on the fact that the UE will use the IMSI as identifier when the TMSI is no longer available. This may happen when the TMSI is deleted by MME or UE due to a timeout. A faster and more active way to catch IMSIs exploits a "fake" base station which acts as a preferred base station in terms of signal strength. Mobile devices typically select base stations emitting the strongest signal. This fake base station can then be used to send an Identity Request message to all mobile devices in the area, which will respond with their IMSIs since they assume that they are connected to a legitimate network which has lost access to the TMSI. In this way, IMSIs of all mobile devices in the area can rapidly be captured.

IMSI catchers were first built for 2G/GSM and later extended to the most recent protocols. The vulnerabilities in the recent protocols allow an adversary to trace the location of users with finer granularity, to enable Denial of Service (DoS) attacks, or to eavesdrop on the communication [14, 15]. General techniques to set up attacks against LTE include traffic capture, jamming, and downgrading to 2G.

To better analyze the robustness of the commercial devices, we create a testbed to test different devices in order to verify how they cope during an IMSI catching attack. We use the Open Air Interface software in order to build an IMSI Catcher on Software Defined Radio (SDR) Universal Software Radio Peripheral (USRP) devices from Ettus research. We test several devices belonging to different vendors, and for each device, we report the success or failure of the IMSI catching attack when a tracking area update (TAU) request is performed, in presence and absence of the jammer module activated.

In Table 5.2 we report results that show the different behaviors that we observed during our experiments for a large variety of mobile terminals. We report here qualitative binary indicators that confirm whether or not attacking a specific user equipment is feasible under which conditions. In Table 5.2 we have considered many heterogeneous devices in order to verify how the operating system could influence the behavior of the device when subject to an IMSI catching attack. At a first glance, it might be conjectured that the OS plays a key role. Indeed, differently from Android devices, iPhone models starting from 7 and including 11 and SE/2020 are more robust to IMSI catching, since they disclose the IMSI only when the attacker performs extensive jamming. As highlighted in Table 5.2 all the considered modems are vulnerable. This is extremely interesting if we consider that the chipsets that we tested have been designed to be compliant with several different versions of the 3GPP standard, from release 8 up to release 15. This proves that, at least at the moment of the test, the operator did not deploy any fix against IMSI catching attacks or fixes have no effect.

5.1.4 Enhanced Privacy Protection in 5G Networks

One of the new entities relevant to 5G authentication is the Subscription Identifier Deconcealing Function (SIDF) used for decryption of a Subscription Concealed Identifier (SUCI) to obtain its long-term identity, namely the SUPI. The SUCI is the encrypted form of the SUPI using the public key of the home network. Thus, a UE permanent identifier, the SUPI, is never sent in clear text over the radio networks in 5G. This feature is considered a major security improvement over prior generations such as 4G. An Elliptic-Curve-Integrated Encryption Scheme (ECIES)-based privacy-preserving identifier containing the concealed SUPI is transmitted.

The 3GPP standard for 5G security specifies two Authentication and Key Agreement (AKA) protocols, Extensible Authentication Protocol AKA' (EAP-AKA') and 5G-AKA.

As it was explained earlier, we tested our jammer against the 5G data session, as we were able to completely tear down the 5G part of the combined/mixed session. We posit that our attack, based on LTE signaling, may still be effective not only in 5G Non Stand Alone (NSA) deployments as we proved, but also against 5G Stand Alone (SA) terminals. Indeed, under the very reasonable assumption that operators, owing to the need to support old phones, will have to run both 4G and 5G deployments in parallel for several years, it should be enough to combine our

Table 5.2 Feasibility of the IMSI catching attack for different mobile devices (left) under different conditions (top).

Model	OS	Modem	3GPP Rel./ LTE Cat.	Without jammer	With jammer
Samsung Galaxy S9	Android 9	Exynos 9810	13/18	✓	✓
Samsung Galaxy A7 2018	Android 10	Exynos 7885	11/12	✓	✓
Samsung Galaxy Note Pro	Android 5	Snapdragon 800	8/4	✓	✓
Nexus 5	Android 6	Snapdragon 800	8/4	✓	✓
Realme X2 Pro	Android 10	Snapdragon X24	13/20	✓	✓
Realme 6	Android 10	Helio G90T	12/13	✓	✓
Xiaomi Redmi Note 7	Android 9	Snapdragon X12	11/12	✓	✓
Xiaomi Mi A1	Android 9	Snapdragon X9	10/7	✓	✓
Huawei Mate 20 Pro	Android 9	HiSilicon Kirin 980	13/21	✓	✓
Huawei P30 Lite	Android 9	HiSilicon Kirin 710	11/12	✓	✓
Huawei P8 Lite	Android 7	HiSilicon Kirin 655	10/6	✓	✓
Asus Zenfone 2	Android 5	Intel XMM 7260	8/4	✓	✓
Vodafone Smart Ultra 6	Android 5	Snapdragon X5	8/4	✓	✓
Orange Neva 80	Android 6	Snapdragon X8	10/6	✓	✓
ZTE Blade v8 lite	Android 7	MediaTek MT6750	10/6	✓	✓
ZTE Blade A910	Android 6	MediaTek MT6735	8/4	✓	✓
ZTE Blade A452	Android 5	MediaTek MT6735	8/4	✓	✓
iPhone 11	iOS 13/14	Intel XMM 7660	13/18	×	✓
iPhone SE 2020	iOS 14	Intel XMM 7660	13/18	×	✓
iPhone XS	iOS 13	Intel XMM 7560	12/16	×	✓
iPhone 8	iOS 13	Intel XMM 7480	12/16	×	✓
iPhone 7	iOS 13	Intel XMM 7360	11/9	✓	✓
iPhone SE	iOS 12/14	Qualcomm MDM9625M	8/4	✓	✓

(Continued)

Table 5.2 (Continued)

Model	OS	Modem	3GPP Rel./ LTE Cat.	Without jammer	With jammer
iPhone 5S	iOS 12	Qualcomm MDM9615M	8/3	✓	✓
Huawei E3272 USB Stick	-	HiSilicon Balong 710	8/4	✓	✓
Huawei E392 USB Stick	-	Qualcomm MDM9200	8/3	✓	✓
iPhone 12	iOS 14	Snapdragon X55	15/22	✓	✓
Xiaomi Mi 10	Android 10	Snapdragon X55	15/22	✓	✓
Oppo Reno	Android 10	Snapdragon X50	N/A	✓	✓
Quectel AG550Q	Android custom	AG215S	N/A	✓	✓

We note that the attack has performed with success, in both 4G and 5G devices, especially when the jammer is enabled.

LTE-based attack with a thorough jamming of the 5G spectrum, which forces the UE to downgrade to LTE connectivity. Consequently, the approach provided in 3GPP standard protocol that improves the 5G security authentication phase will continue to expose some weaknesses up to the user device enable the downgrade support.

5.1.5 Location Privacy Algorithms

The aforementioned new 5G location enhanced capabilities create enormous opportunities for new LBS. These capabilities have great potential for the 5G operators that can exploit them to offer new useful services to their customers. Examples of LBS include network planning, maintenance support, flow monitoring, tracking systems, location-based advertising, and social networking services. However, significant privacy concerns are raised by LBS. This section faces the privacy aspects related to LBS in 5G architecture and presents preliminary technical solutions to process data in compliance with users' privacy rights.

An LBS uses location data, which are related to the smartphone and/or mobile device in order to deliver services to the users. Consequently, privacy attacks on the cellular network may lead to privacy issues for the user's location. Indeed, from the user's point of view, location data transmitted over the network are highly sensitive. If these data are revealed, the location, the trajectory, and the identity of the users can be leaked.

Privacy threat happens when an attacker can create an association between identity, request content and location of a user [16]. This information can be possibly obtained from location-based requests to LBS, specifically when the background knowledge is available.

In the case of LBS, this sensitive association can be possibly derived from location-based requests issued to service providers. The identity and the private information of a single user can be derived also from requests issued by a group of users as well as from available background knowledge. Here, one of the challenges is to keep data simply usable by third-party stakeholders that can provide LBSs on a specific localization platform without raising any privacy issues.

To mitigate the privacy issues, the implemented system needs to include functional blocks dedicated to location privacy. The most common platform functions are reported below:

- Sanitization: This function implements the process of removing user sensitive information from stored location data;
- k-Anonymity: This function implements a data process to produce data output where a user cannot be distinguished from at least the other $k-1$ individuals;
- Obfuscation: This function implements techniques that aim at blurring or perturbing the location information;
- Policy definition: This function implements the setting of the privacy policy. It describes a fine-grained sanitization policy developed for private data, to facilitate their release to the public, and a sanitization tool that applies the policy;
- Result aggregation: The privacy of a user/device is further protected by using a "hiding in a crowd" approach 3rd party can query any features which interest them but they only receive aggregate responses (counts, histograms, etc.) to address data query correlations.

In this chapter we focus on two of these functionalities, k-Anonymity and Result Aggregation. Specifically, this document shows the descriptions and the performances of the designed algorithms.

Regarding this topic, the main body of research includes approaches that are based on the notion of k-Anonymity, which has been originally proposed by Samarati and Sweeney [17] in the context of relational data. To satisfy k-Anonymity in LBSs, the most widely adopted anonymization strategies are:

- Dummy location approach [18]
- Data/spatial cloaking technology [19]
- Historical/trajectory approaches [20]

Dummy location approach: The dummy location approach generates multiple dummy locations and integrates the users' real locations into the dummy ones and sends them to the service provider for privacy protection.

Data/spatial cloaking technology: Data-dependent cloaking strategies formulate the region of anonymity based on the actual location of each user in the system and their distance from the location of the request. Specifically, these strategies retrieve the k-1 nearest neighbors of the requester and generate a region that includes all the k users. In [21] an anonymous region constructing algorithm based on kd-tree [22] is proposed. It works in densely and sparsely populated regions.

Historical/trajectory approaches: The cloaking approaches, presented before, do not consider the user story requests, which can be used by an attacker to extract the user's private information. While historical approaches to k–Anonymity keep track of the movement history of each user in the system and utilize this information when building the anonymity regions for the user requests. Finally, trajectory k–Anonymity approaches consider the analysis of future requests. Both these approaches are appropriate for services in which the current position of the user has to be communicated to the service provider for as long as the user travels.

5.1.6 Location Privacy Considered Model

To show how it is possible to improve the robustness of the LBS architecture to the privacy threat, we consider the model with an untrusted LBS server, that starting from the user's queries with position and content requests, which can track users or release their personal data to third parties. Consider, for example, a service request originating from the house of a user. The request contains sufficient information to extract the user identity, even though any other identification data are not present in the requested content. Indeed, the untrusted LBS can map the exact coordinates that are part of the user request to a publicly available house owner registry.

The considered process to improve privacy robustness in LBSs is based on three different phases. We consider a population of users who are supported by the network infrastructure. For each user, the system performs a location update to the platform data storage. A set of 3rd party LBSs are subscribed to the domain. In the phase 1, the incoming request arrives from the users. The role of the privacy module is to filter the incoming user requests (phase 2) and to produce anonymous counterparts that can be safely forwarded to the LBS (phase 3). To produce the anonymous counterpart, the privacy module has to apply functionalities presented before.

5.1.7 Location Privacy Tested Approach

In this subsection, we report an algorithm to create k-Anonymity regions when kd-tree data structure is considered together with the testing results. To exhibit the test results, we use real data containing the position of a population of operator users that move in a region of 400 square. We first present some concepts related to the entropy evaluation for a group of k neighbor users in a population of $2k$ users and then introduce the procedure of the tested algorithm. The anonymous entropy method is considered to select users who are evenly distributed with real users (u_{real}), including the requested content. The goal of the anonymous entropy method on distance is as follows: if the sum of distances between $k - 1$ users and the real user among m user groups are equal, the user group which is evenly distributed is selected. Otherwise, if the total distances are not equal, the user group with a larger distance is selected. In order to achieve the above goal, entropy is used to select a user group on distance. Here, the weight of the neighbor user u_i in the nth user group is denoted as α_{ni}, that is,

$$\alpha_{ni} = d(u_{\text{real}}, u_i) / \sum_{j=1}^{2k} d(u_{\text{real}}, u_j) \ (i = 1, \ldots, k - 1), \tag{5.1}$$

where $d(u_i, u_j)$ denotes the Euclidean distance between u_i and u_j.

Additionally, the requested content among users is considered. A weight is assigned to the requested content in the user group based on the number of request types and the distribution in the user group, that is

$$\beta_n = 2\text{bool}(u_i^c, u_j^c) / k(k - 1) \ (i, j = 1, \ldots k, i \neq j). \tag{5.2}$$

Taking into account the distance between users in the user group and the distribution of the request content jointly, the anonymous entropy for the nth user group is defined as the sum of the entropy in terms of distance and the difference in request content in the nth user group, denoted by,

$$H_n = \sum_{i=1}^{k-1} \alpha_{ni} \log_{10} \alpha_{ni} + \beta_n. \tag{5.3}$$

Finally, the entropy H_n is used to measure the uncertainty of a group of users the bigger the value of the entropy, the more uncertain the group. Formally, it is defined as where the probability that the user is identified. In many cases, the entropy is used to evaluate the anonymity of anonymous regions. Here, we adopt the same evaluation criteria as well.

The defined method works similar to [21], after collecting the LBS requests sent by users, the module anonymously processes the location privacy, identity, request content, and other information according to the requirements of the users. The procedure is as follows:

1. Construct the kd-tree based on the region where the requesting user is located.
2. The nearest neighbor users are searched on the kd-tree, and the anonymity is processed according to the algorithm.

The selected algorithm works in this mode: Suppose that there are $2k$ users in a region, $k - 1$ users are selected randomly to form a user group U_i with real users. The process is repeated m times, and m user groups are formed, where m is defined by the user according to their privacy requirements. In sparsely populated regions, considering the historical records and geographical distribution, the algorithm achieves k-Anonymity by selecting dummy users with high historical query probability, relatively uniform geographical distribution, and large difference in request contents.

In this subsection, we evaluate the performance of our scheme via extensive experiments. For these experiments, we use the real dataset anticipated before and we generate up to 1000 queries originating from random users, we consider 15 different types of request contents. System performances are monitored by the evaluation of the anonymity spatial region computes as the minimum bounding rectangle (MBR) enclosure that contains the k users in the group of the requester. Better performances are obtained when for a certain entropy we have a more small region size. We generate MBR and compare the proposed algorithm with the quad-tree algorithm used in Casper [23]. To evaluate the performance we use entropy results and the area of the anonymous region at different k-value.

Figure 5.1 shows the comparison of anonymity between the proposed algorithm and the Casper algorithm under different k values, in terms of entropy (a) and anonymous region area (b). The entropy perceived in the proposed approach presents a slope gradient with respect to the Casper method. At the same time, the area of the anonymous region formed by quad-tree algorithm, used in Casper, increases with the rise of the k value. We note that the number of people in the region is 1000 here. Different from the proposed method, in Casper with the increase of k value, the area of anonymous regions becomes increasingly large. Quad-tree algorithm do not take full account of the location relationships of the neighboring users. In particular, when it extends to a high level, each expansion leads to a larger area increase, generating redundant space.

5.2 Location Security

Location data in 5G networks will rely on metrics extracted from the New Radio (NR) uplink and downlink reference signals (3GPP-technologies) as well as on non-3GPP technologies such as, Global Navigation Satellite System (GNSS), Terrestrial Beacon Systems (TBS), WLAN, and other sensors. In any case, the security requirements for localization techniques must include the capability of identifying and possibly mitigating any kind of deviation from the true locations. Location data are highly vulnerable to both data-level and signal-level spoofing and meaconing attacks caused by malicious intruders. In order to face the threats

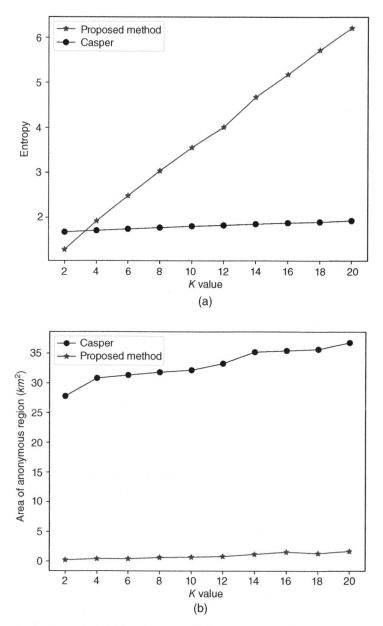

Figure 5.1 Comparison of anonymity between the proposed algorithm and the Casper algorithm under different *k* values, in terms of entropy (a) and anonymous region area (b).

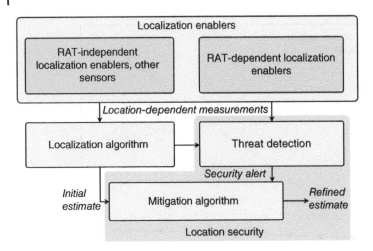

Figure 5.2 High-level description of the location security function and its connections with the localization function and heterogeneous sensors.

mentioned above, a generic location security function should contain a preliminary stage responsible for the detection of an attack and a second stage (possibly shared with the location function) aimed at handling the fake measurements generated by the malicious actor. A high-level description of the designed location security function is shown in Figure 5.2.

As shown in Figure 5.2, the location security function must be able to manage and perform data fusion of measurements provided by heterogeneous sensors.

5.2.1 Location Security in 4G/5G Networks

5G Location services mainly rely on the (possibly joint) exploitation of suitable positioning measurements such as the Time of Arrival (ToA), Time Difference of Arrival (TDoA), Angle of Arrival (AoA), and Received Signal Strength (RSS) from 5G networks or other integrated non-3GPP technologies. The localization process may occur at a network level (network-centric approach), where the computation of the UE position is performed by the network and transmitted to the latter, or at a user level (user-centric solution), where the UE collects information from the network and uses it to determine its position. In both cases, the location measurements can be combined with the information provided by other sources, for instance, satellite positioning systems or other UEs in the proximity of that under consideration to enhance the localization accuracy. In this highly connected context, location security becomes of primary concern, especially in applications related to safety and liability. As a matter of fact, due to the large number of stages interacting toward the position estimation, the exposure to attacks is high indeed. Otherwise stated, there exists a plethora of points of the distributed communication system or, equivalently, of the processing chain that can be subject to malicious actions. Therefore, new solutions aimed at preserving data integrity are strongly desirable. Among the possible threats to location security, here we focus on intentional interference, which involves hostile platforms that target the UE and AN receivers in order to:

- reduce the signal-to-noise ratio (noise-like jamming)
- inject false or erroneous information (spoofing/meaconing).

Hostile platforms that perform a DoS attack by transmitting high-power interfering signals to induce the disruption of the receiver functionalities (see Figure 5.3) belong to the first class of threats and are referred to as noise-like jammers (NLJs). Signals transmitted by NLJs blend into the receiver's thermal noise with an increase of the noise power spectral density within the receiver bandwidth.

As for the second class of attacks, spoofing/meaconing threats intercept the positioning messages exchanged by two legitimate actors and alter them by synthetizing counterfeit information. These erroneous messages would prevent the location service from providing reliable position estimates (see Figure 5.4)

Figure 5.3 Jamming operating scenario.

5.2.2 Threat Models and Bounds

The scenario considered is very classical, where an end-device infers its position by means of suitable measurements taken from a set of reference *anchor stations*

Figure 5.4 Spoofing/meaconing operating scenario.

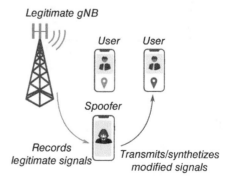

whose position is assumed known. A *location spoofing attack* can be technically performed in several ways: by altering the measurement process so that the reference anchor station is perceived as closer (or farther, or shifted) from its authentic place, or by deploying a rogue station claiming to be a legitimate one but placed in a different position, or by corrupting the control system which provides the legitimate anchors' positions.

The threat model proposed aims to abstract from the specific details of each attack and instead has the ambition to provide a formal reference model common to all the above specific cases. The intuitive idea is that a location attack occurs when the attacker is capable to *associate an anchor's position to an observable not representative of the claimed position*, being irrelevant whether this is obtained by tampering with the measurements or by spoofing the claimed position. In the following, we formalize this notion.

5.2.2.1 Formal Model

Consider a localization network consisting of N_b anchors for inferring the location of an agent, which is at $\mathbf{p}$.[1] The ith anchor is at position $\mathbf{p}_i$, and $\mathbf{p}_b = [\mathbf{p}_1, \mathbf{p}_2, \dots, \mathbf{p}_{N_b}]$. The agent location is inferred based on measurements of signals communicated between each anchor and agent. In particular, the measurement vector is $\mathbf{z} = [z_1, z_2, \dots, z_{N_b}]$, where z_i is measured between the ith anchor and the agent.

An example is illustrated in Figure 5.5. The localization algorithm exploits $\mathbf{z}$ together with the information about the anchors' positions. Each measurement depends on the true anchor and agent positions according to a measurement model, e.g.

$$z_i = z_0(\mathbf{p}, \mathbf{p}_i) + n_i, \tag{5.4}$$

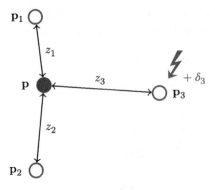

Figure 5.5 Example scenario with $N_b = 3$ anchors and one agent in the presence of a spoofing attack against the third anchor.

---

1 We consider $\mathbf{p} \in \mathbb{R}^2$.

where $n_i \sim \mathcal{N}(0, \sigma_i^2)$ and the measurements from different anchors are independent. Example cases are when the measurement is a timing, angle, or power measurement, and we know the signal speed and the anchors' position.

If we model the agent position as a deterministic but unknown parameter and the anchor positions as a deterministic and known parameter, the likelihood function for the vector $\mathbf{p}$ is

$$f(\mathbf{z}, \mathbf{p}, \mathbf{p}_b) = \prod_{i=1}^{N_b} f(z_i, \mathbf{p}, \mathbf{p}_i), \tag{5.5}$$

where each $f(z_i, \mathbf{p}, \mathbf{p}_i)$ is obtained according to the measurement model in (5.4).

If the likelihood function is known, the maximum likelihood (ML) estimator is the optimal solution, as it achieves the Cramér-Rao bound (CRB) asymptotically in the high signal-to-noise ratio (SNR) regimes. The ML estimator is unbiased, i.e. $\mathbb{E}\{\hat{\mathbf{p}}\} = \mathbf{p}$, where $\hat{\mathbf{p}}$ is the estimate of $\mathbf{p}$. In most cases, the likelihood function is generally unknown, as the parameters of the measurement distribution can be unknown (or, at most, partially known). In such practical cases, sub-optimal estimators are considered, e.g. using the well-known trilateration algorithm or the least square algorithm.

5.2.2.2 Error Model for the Spoofing Attack

In the presence of a spoofing attack, where the anchor positions are tampered, the main effect is that the measurement z_i is taken with respect to the true anchor at $\mathbf{p}_i$, and therefore follows the true measurement model $z_0(\mathbf{p}, \mathbf{p}_i) + n_i$. Nevertheless, as the information about the anchor position is tampered, i.e. the information on $\mathbf{p}_i$ is biased as $\mathbf{p}_i + \delta_i$, where δ_i the bias, if there is no detection or awareness of such a tampering attack, the localization algorithm will estimate the agent position according to an incorrect measurement model, i.e. $z_0(\mathbf{p}, \mathbf{p}_i + \delta_i) + n_i$. The effect of such an incorrect measurement model on localization accuracy depends on several system parameters and the estimator itself. Different estimators will generally be less or more robust to this type of attack.

In the case of an ML estimator, the position estimate under attack will be

$$\hat{\mathbf{p}}_{sp} = \arg \max_{\tilde{\mathbf{p}}} f(\mathbf{z}, \tilde{\mathbf{p}}, \mathbf{p}_i + \delta_i). \tag{5.6}$$

Note that for $\delta_i \neq \mathbf{0}$ for some i, the ML estimator is biased, i.e. $\mathbb{E}\{\hat{\mathbf{p}}_{sp}\} \neq \mathbf{p}$. We define the spoofing error as $\mathbf{e}_{sp} = \hat{\mathbf{p}}_{sp} - \mathbf{p}$. Let us now consider the following system of N_b equations with respect to $\check{\mathbf{p}}$

$$z_0(\check{\mathbf{p}}, \mathbf{p}_i + \delta_i) = z_0(\mathbf{p}, \mathbf{p}_i) \quad \forall i = 1, 2, \ldots, N_b. \tag{5.7}$$

If there exists a solution to (5.7), such vector $\check{\mathbf{p}}$ would be the position of the agent in the case the true position of the ith anchor would be $\mathbf{p}_i + \delta_i$ for each $i = 1, 2, \ldots, N_b$ and the measurement between the anchor and the ith anchor would have the expected value z_i. In such a case, i.e. in the absence of any spoofing, an ML estimator for the case with an agent at $\check{\mathbf{p}}$ and the anchors $\mathbf{p}_i + \delta_i$ would solve the equivalent problem as in (5.6) as an unbiased estimator. Then, $\mathbb{E}\{\hat{\mathbf{p}}\} = \check{\mathbf{p}}$. It follows that, being this the identical problem as (5.6) we have

$$\mathbb{E}\{\hat{\mathbf{e}}_{\text{sp}}\} = \check{\mathbf{p}} - \mathbf{p}. \tag{5.8}$$

Note that (5.8) is valid for any estimator that is unbiased in the absence of an attack, i.e. $\mathbb{E}\{\hat{\mathbf{p}}\,|\,\delta = \mathbf{0}\} = \mathbf{p}$ and that is based on a measurement model as in (5.4). If $\check{\mathbf{p}}$ does not exist, i.e. the system of N_b equations in (5.7) has no solution, then the error will depend on the specific localization algorithm and the measurement model.

5.2.2.3 Threat Model Example Case Study: Range-Based Localization Using RSSI

As an example, we here focus on the range-based localization using received signal strength indicator (RSSI). In this case, each anchor transmits with power P_T. The signal propagates in fading channel where the fading is modeled as a log-normal random variable $n_i \sim \mathcal{N}(0, \sigma^2)$. Thus, the power received at the agent from the ith anchor is

$$z_i = 10 \log_{10} \frac{P_T}{d_i^\eta} + n_i, \tag{5.9}$$

where $d_i = \|\mathbf{p} - \mathbf{p}_i\|$ is the true distance between the ith anchor and the agent, η is the path-loss exponent, and $n_i \sim \mathcal{N}(0, \sigma^2)$ are statistically independent.

Then, the received power is considered as the measurement for each anchor-agent link, and the power-distance law gives the measurement model as a function of the positions through the path-loss exponent.

5.2.2.4 Error Bound Under Spoofing Attack

Considering the measurement model $f(\mathbf{z}_i, \mathbf{p})$ for the observation z_i and unknown deterministic parameter vector $\mathbf{p}$. Let $\hat{\mathbf{p}}$ be any unbiased estimate of $\mathbf{p}$ given $\mathbf{p}_i$. Based on the information inequality, which gives a lower bound on the mean squared error (MSE) of estimators, we have

$$\mathbb{E}\{\|\hat{\mathbf{p}} - \mathbf{p})\|^2\} \geq \text{tr}\{\mathbf{J}_{\mathbf{p}}^{-1}\}, \tag{5.10}$$

where $\mathbf{J}_{\mathbf{p}}$ is the Fisher information matrix for the parameter vector $\mathbf{p}$ and $\text{tr}\{\mathbf{J}_{\mathbf{p}}^{-1}\}$ is called the squared position error bound (SPEB) [24].

As we have discussed earlier, an estimator $\hat{\mathbf{p}}$ that is unbiased in the absence of a tampering attack, i.e. $\mathbb{E}\{\hat{\mathbf{p}}\,|\,\delta = \mathbf{0}\}$, becomes biased when $\delta \neq \mathbf{0}$ due to the incorrect measurement model. In such a case, $\mathbb{E}\{\hat{\mathbf{p}}\,|\,\delta \neq \mathbf{0}\} = \mathbf{p} + \mathbf{e}_{\text{sp}}$, where $\mathbf{e}_{\text{sp}}$ is the bias due to the tampering attack.

The information inequality on the mean squared error of such a biased estimators should take into account the bias $\mathbf{e}_{sp}$. In particular, we define

$$\mathbf{\Psi}_{\hat{\mathbf{p}},\delta} = \frac{\partial \mathbb{E}\{\hat{\mathbf{p}}\,|\,\delta \neq \mathbf{0}\}}{\partial \mathbf{p}}, \tag{5.11}$$

and we derive the SPEB for a biased estimator $\hat{\mathbf{p}}$ as

$$\mathbb{E}\{\|\hat{\mathbf{p}} - \mathbf{p}\|^2\,|\,\delta \neq \mathbf{0}\} = \mathrm{tr}\left\{\mathbf{\Psi}_{\hat{\mathbf{p}},\delta}\mathbf{J}_{\mathbf{p}}^{-1}\mathbf{\Psi}_{\hat{\mathbf{p}},\delta}^{\mathrm{T}}\right\}. \tag{5.12}$$

5.2.2.5 Case Study

In this section, we evaluate the effects of tampering on location estimation using simulation results. We consider a network on $N_b = 3$ anchors uniformly distributed on a circumference of radius $r = 1\,\mathrm{km}$. We consider the agent as uniformly distributed within a squared area of 1 by 1 km. RSSI-based localization is considered following the measurement model in (5.9) with σ varying from 0.1 to 10, and $\eta = 2$. The spoofing is simulated considering a constant value $\delta_i = [\delta, \delta]$ equal for all the spoofed anchors. We consider the case with a single spoofed anchor and two spoofed anchors. Location estimation is performed with the least square algorithm, which is equivalent to the MLE when σ is constant and the underlying distribution is Gaussian.

Figure 5.6 shows the SPEB and MSE varying δ when a single or two anchors are spoofed. The second spoofed anchor increases both the MSE and the SPEB.

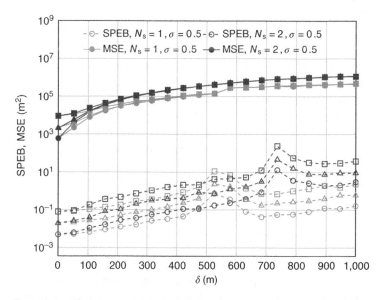

Figure 5.6 SPEB (dashed) and MSE (solid) varying δ, with $\sigma = 0.5$ (circles), $\sigma = 1$ (triangles), and $\sigma = 2$ (squares); single spoofed anchor (light gray) and two spoofed anchors (dark gray).

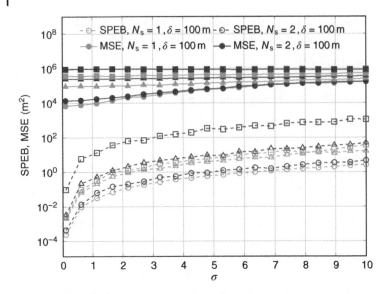

Figure 5.7 SPEB (dashed) and MSE (solid) varying σ with $\delta = 100$ m (circles), $\delta = 400$ m (triangles), and $\delta = 800$ m (squares); a single spoofed anchor (light gray) and two spoofed anchors (dark gray).

Note that the value of the MSE with two spoofed anchors and $\delta = 270$ m is comparable to the MSE with a single spoofed anchor with $\delta = 350$ m. As a matter of fact, the value of the bias is the leading parameter and therefore even a single spoofed anchor can impact dramatically the localization performance.

Figure 5.7 shows the SPEB and the MSE as a function of σ for $\delta = 100, 400,$ and 800 m with a single or two spoofed anchors. When the value of δ is above 100 m, the effect of sigma is negligible for any value of σ in the interval considered. Also, when $\delta = 100$ m, the effect of the number of spoofed anchors is much smaller than when $\delta > 100$ m. This fact corroborates what is observed in Figure 5.6 and shows that the measurement noise has a little impact in the presence of spoofing attacks.

Figure 5.8 shows the MSE varying the number of anchors N_b for the case with a single or two spoofed anchors. As it could be expected, the MSE decreases with the number of anchors that are not affected by spoofing. In particular, with $N_b = 8$, the case with a single spoofed anchor is very close to the case without spoofing, meaning that the effect of the spoofing has been mitigated with a greater number of anchors. On the other side, when two anchors are spoofed, even $N_b = 8$ anchors are not sufficient to mitigate completely the effect of the spoofing. These results provide a quantitative indication of the number of non-spoofed anchors required to compensate the bias introduced by the spoofed anchors.

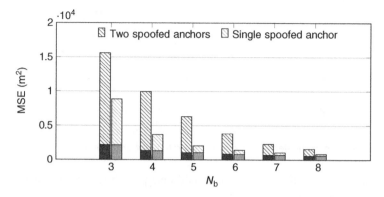

Figure 5.8 MSE for different numbers of anchors in the case of two spoofed anchors (dark gray) and a single spoofed anchor (light gray), with spoofing (dashed) and without spoofing (full).

5.3 3GPP Integrity Support

Aside from high precision positioning, which is one of the main demands in emerging applications, and the aspects of privacy and security, which have been earlier discussed and evaluated in this chapter, high positioning integrity and reliability are other demands for many use-cases.

A positioning system requires to have both location accuracy and reliability, or precision, when computing location estimation. Precision means exactness. In simple terms, precision is how small of a position you put on the map. Precision is a mathematical term, and we express it using latitude/longitude. The more decimal places you have, the more precise it is. Accuracy is where we fix a position relative to ground truth. It is how close the position is to reality. This is where we talk about horizontal positioning error. Finally, integrity requires the evaluation of both accuracy and reliability. Figure 5.9 illustrates the definitions of accuracy, precision, validity, reliability, integrity, and confidence level, which have been extensively used in the literature.

Integrity is the measure of trust that can be placed on the correctness of information supplied by a navigation system. Integrity includes the ability of a system to provide timely warnings to user receivers in case of failure. This topic plays an important role in use-cases in which errors can lead to serious consequences such as wrong legal decisions or wrong charge computation. If positioning integrity is considered to be sufficient (i.e. reliable and trustworthy, no residual error), the application using the positioning information can operate according to its standard operating procedures and in accordance with application safety requirements. However, if the positioning integrity is assumed to be insufficient

X Unreliable, imprecise ✓ Reliable, precise   X Unreliable, imprecise ✓ Reliable, precise
X Inaccurate, invalid   X Inaccurate, invalid   ✓ Accurate, valid   ✓ Accurate, valid
X Inconsistent   ✓ Consistent   X Inconsistent   ✓ Consistent
X Low confidence level ✓ High confidence level X Low confidence level   ✓ High confidence level
X No integrity   X Lack of integrity   X Lack of integrity   ✓ High integrity

Figure 5.9 Illustration of the definitions of accuracy, reliability, and integrity.

(i.e. large residual error), the application should take predefined precautionary actions to prevent negative outcomes [25].

The topic of integrity started to be studied in 3GPP since Rel-17. The work in Rel-17 was mainly concentrated on GNSS integrity support; however, in Rel-18 the intention is to expand the support to RAT-dependent positioning methods as well. The 3GPP positioning integrity support enables UEs to access the correctness of information and expected position estimation accuracy, and relate it to safety margins in order to determine availability of the positioning estimates to safety critical applications and even obtain these parameters via broadcast.

The study and support on this topic in 3GPP are handled by identifying positioning integrity KPIs and relevant use cases, and the error sources, threat models, occurrence rates, and failure modes requiring positioning integrity validation and reporting. There are some integrity KPIs based and extracted from the satellite navigation area which are applied to the 5G integrity positioning support. Below these KPIs are summarized:

- Alert Limit (AL) is the largest allowable error for safe operation by the application
- Time to Alert (TTA) is the amount of time during which the position error can be higher than the alert limit before an alarm is triggered.
- Target Integrity Risk (TIR) is the probability that the position error is larger than the alert limit without an alarm being triggered.

The Protection Level (PL), which is the distance within which the true position is contained within the probability of the TIR, can be determined with the availability of the above KPIs. Figure 5.10 provides the fundamentals of positioning integrity KPI definition and how they can be used to form a position error CDF to compute the PL. In the time when PL exceeds the AL for TTL, then the system would send alarm indicating that the reliability is not sufficient and available for the positioning system. Depending on the capabilities of the network and the UE, it is possible to support operation at different integrity levels. There is a possibility to set integrity thresholds for positioning quality of service. Integrity levels can be

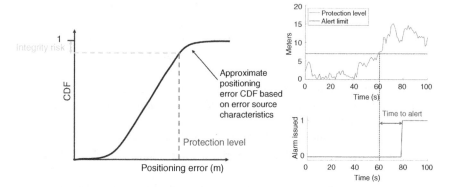

Figure 5.10 Positioning integrity KPI fundamentals [25].

defined for the overall positioning system, and these levels can be determined or requested either before or during positioning procedures [25].

The integrity support for RAT-dependent positioning methods is being studied and evaluated in Rel-18 and would be part of the enhancements in 5G-Advanced positioning support. In case of GNSS integrity, the following feared events and error sources have been already summarized in Rel-17 [26]. They can be categorized as in feared events in the GNSS assistance data, feared events during positioning data transmission, GNSS feared events, i.e. coming from satellite, atmospheric or local environment (such as multipath, interference, and spoofing), UE feared events, and LMF feared events.

References

1 TR 38.855. 3rd Generation Partnership Project (3GPP), Technical Report Group Radio Access Network; Study on NR positioning support, March 2019. Release 16.

2 C. Bettini, S. Mascetti, D. Freni, X. Wang, and S. Jajodia. Privacy and anonymity in location data management. *Privacy-Aware Knowledge Discovery: Novel Applications and New Techniques*, 12 2010.

3 R. L. Villars, C. W. Olofson, and M. Eastwood. Big data: What it is and why you should care, 2011. URL https://www.admin-magazine.com %2FHPC%2Fcontent%2Fdownload%2F5604%2F49345%2Ffile%2FIDC_ BigData_whitepaper_final.pdf&psig=AOvVaw2zPf2EFpVs0d32GFXj1PG5& ust=1666865501137634. Last Accessed: May 8, 2023.

4 Z. Wang, G. Wei, Y. Zhan, and Y. Sun. Big data in telecommunication operators: Data, platform and practices. *Journal of Communications and Information Networks*, 2(3):78–91, 2017.

5 R. Di Taranto, S. Muppirisetty, R. Raulefs, D. Slock, T. Svensson, and H. Wymeersch. Location-aware communications for 5G networks: How location information can improve scalability, latency, and robustness of 5G. *IEEE Signal Processing Magazine*, 31(6):102–112, 2014.

6 M. Koivisto, A. Hakkarainen, M. Costa, P. Kela, K. Leppanen, and M. Valkama. High-efficiency device positioning and location-aware communications in dense 5G networks. *IEEE Communications Magazine*, 55(8):188–195, 2017.

7 H. Khan, B. Dowling, and K. M. Martin. Identity confidentiality in 5G mobile telephony systems, pages 120–142. Springer International Publishing, Cham, 2018.

8 TR 33.501. 3rd Generation Partnership Project (3GPP), Technical Specification (TS); Security architecture and procedures for 5G System, August 2018.

9 TS 33.401. 3rd Generation Partnership Project (3GPP), 3GPP system architecture evolution (SAE)–security architecture, September 2022. Release 17.

10 A. Lilly. IMSI catchers: Hacking mobile communications. *Network Security*, 2017(2):5–7, 2017.

11 R. Olimid and S. Mjølsnes. On low-cost privacy exposure attacks in LTE mobile communication. *Proceedings of the Romanian Academy - Series A: Mathematics, Physics, Technical Sciences, Information Science*, 18:361–370, 2017.

12 Z. Li, W. Wang, C. Wilson, J. Chen, C. Qian, T. Jung, L. Zhang, K. Liu, X. Li, and Y. Liu. FBS-Radar: Uncovering fake base stations at scale in the wild. February 2017.

13 B. Hong, S. Bae, and Y. Kim. GUTI reallocation demystified: Cellular location tracking with changing temporary identifier. In *Proceedings of Network and Distributed Systems Security (NDSS) Symposium*, San Diego, CA, USA, February 2018.

14 R. P. Jover. LTE security, protocol exploits and location tracking experimentation with low-cost software radio, 2016.

15 D. Rupprecht, K. Jansen, and C. Pöpper. Putting LTE security functions to the test: A framework to evaluate implementation correctness. In *Proceedings of the USENIX Conference on Offensive Technologies*, page 40–51, Austin, TX, USA, 2016.

16 F. Gustafsson and F. Gunnarsson. Mobile positioning using wireless networks: Possibilities and fundamental limitations based on available wireless network measurements. *IEEE Signal Processing Magazine*, 22(4):41–53, 2005.

17 P. Samarati. Protecting respondents identities in microdata release. *IEEE Transactions on Knowledge and Data Engineering*, 13(6):1010–1027, 2001.

18 D. Wu, Y. Zhang, and Y. Liu. Dummy location selection scheme for k-anonymity in location based services. In *2017 IEEE Trustcom/BigDataSE/ICESS*, pages 441–448, Sydney, Australia, 2017.

19 O. Abul, F. Bonchi, and M. Nanni. Never walk alone: Uncertainty for anonymity in moving objects databases. In *Proceedings of IEEE International Conference on Data Engineering*, pages 376–385, 2008.

20 T. Xu and Y. Cai. Exploring historical location data for anonymity preservation in location-based services. In *Proceedings of the IEEE Conference on Computer Communications (INFOCOM)*, pages 547–555, Phoenix, AZ, USA, May 2008.

21 L. Ni, F. Tian, Q. Ni, Y. Yan, and J. Zhang. An anonymous entropy-based location privacy protection scheme in mobile social networks. *EURASIP Journal on Wireless Communications and Networking*, 2019:93, 2019.

22 M. J. Dürst. The design and analysis of spatial data structures. Applications of spatial data structures: Computer graphics, image processing, and GIS. *The Visual Computer*, 7(2):170–170, 1991.

23 M. F. Mokbel, C.-Y. Chow, and W. G. Aref. The new Casper: A privacy-aware location-based database server. In *IEEE International Conference on Data Engineering*, pages 1499–1500, 2007.

24 M. Z. Win, Y. Shen, and W. Dai. A theoretical foundation of network localization and navigation. *Proceedings of the IEEE*, 106(7):1136–1165, 2018.

25 F. Gunnarsson and R. Shreevastav. 3GPP GNSS positioning and integrity: The latest trials and developments. *Ericsson Blogpost*, 2022.

26 TR 38.857. 3rd Generation Partnership Project (3GPP), Technical Report Group Radio Access Network; Study on NR positioning enhancements, March 2021. Release 17.

Part II

Location-based Analytics and New Services

6

Location and Analytics for Verticals

Gürkan Solmaz[1], Raquel Barco[2], Stefania Bartoletti[3], Andrea Conti[4], Nicolò Decarli[5], Yannis Filippas[6], Andrea Giani[4], Emil J. Khatib[2], Oluwatayo Y. Kolawole[7], Tomasz Mach[7], Barbara M. Masini[5] and Athina Ropodi[6]

[1]NEC Laboratories Europe, Heidelberg, Germany
[2]Telecommunication Research Institute (TELMA), University of Malaga, E.T.S.I. de Telecomunicación, Málaga, Spain
[3]Department of Electronic Engineering and CNIT, University of Rome Tor Vergata, Italy
[4]Department of Engineering and CNIT, University of Ferrara, Ferrara, Italy
[5]National Research Council – Institute of Electronics, Computer and Telecommunication Engineering and WiLab-CNIT, Bologna, Italy
[6]Incelligent P.C., Athens, Greece
[7]Communications Research, Samsung Electronics R&D Institute UK, Staines-upon-Thames, England, United Kingdom

This chapter investigates the location-based analytics for vertical applications. Section 6.2 discuss people-centric data analytics, providing the reader with example analytics for crowd mobility as well as the use of location and location-based analytics for the road safety applications.

6.1 People-Centric Analytics

6.1.1 Crowd Mobility Analytics

6.1.1.1 Introduction and Related Work

The understanding of the crowd mobility in environments such as smart cities would enable improvements in the services for various domains such as tourism, event management, retail, and epidemic monitoring. Therefore, this section describes the data analytics for crowd monitoring and tracing. As a low-cost and non-privacy-invasive alternative to deployment of vast cameras, pervasive application through wireless networks and wireless sensing data is considered. First, this section introduces the machine-learning-based crowd estimation

Positioning and Location-based Analytics in 5G and Beyond, First Edition.
Edited by Stefania Bartoletti and Nicola Blefari Melazzi.

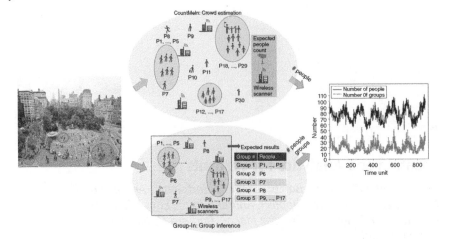

Figure 6.1 Crowd monitoring and group inference for smart cities through wireless scanning. The outputs of the analytics are the number of people and number of people groups in the environment at given time intervals. Source: chensiyuan/Wikimedia Commons/CC BY-SA 3.0.

with examples using wireless scanners and a limited number of stereoscopic cameras. Second, it describes machine-learning-based people group inference using wireless devices such as Bluetooth scanners. Figure 6.1 illustrates the crowd estimation and group inferences for different urban environments. The goal is to accurately estimate the number of people and number of people groups [1–5]. People groups are defined based on their physical closeness that may be inferred by wireless fingerprints. Table 6.1 lists the acronyms used in this Chapter.

Crowd estimation is one of the major topics for smart city applications in general. The aim of the crowd estimation analytics is to provide long-term and high accuracy estimation in smart cities, more particularly, in areas of interest in the cities such as shopping streets, train stations, city squares, and other hot-spot locations in urban areas. As an example, the crowd estimation analytics presented in this section considers two sensing modalities: (i) Wi-Fi RSSI from the urban environment and (ii) Stereoscopic cameras [6].

In [5], the setup of Wi-Fi and stereoscopic cameras is tested in a central train station where the setup considered multiple stereoscopic cameras in the station always actively collecting measurements of people counts. These people counts are highly accurate, and thus they have been used to correlate crowd estimations using Wi-Fi fingerprints in real-time. Moreover, the correlations have been applied to the nearby areas, and they can be useful in a very close vicinity, whereas applying to another area expectedly decrease the crowd estimation accuracy. Therefore, machine-learning-based systems are employed to train Wi-Fi and camera correlations [2] in different environments of the city, in short training periods, and

Table 6.1 List of acronyms.

Acronym	Definition
3GPP	3rd Generation Partnership Project
5G	5th generation
5GAA	5G Automotive Association
ADE	Average displacement error
AI	Artificial intelligence
AoD	Angle of departure
CAM	Cooperative awareness message
C-ITS	Cooperative intelligent transport systems
C-V2X	Cellular vehicle-to-everything
DB	Database
DEN	Decentralized environmental notification
DL	Deep learning
EGNOS	European geostationary navigation overlay service
ETSI	European Telecommunication Standards Institute
Euro NCAP	European new car assessment programme
FDE	Final displacement error
GAN	Generative adversarial network
GNSS	Global navigation satellite system
GRU	Gated recurrent unit
HCS	Highly-connected subgraphs
ITS	Intelligent transport systems
ITS-S	Intelligent transport systems station
KPI	Key performance indicator
LDM	Local dynamic map
LSTM	Long short-term memory
LTE	Long-term evolution
MDE	Mean distance error
ML	Machine learning
MLP	Multi-layer perception
MSE	Mean squared error
OSI	Open System Interconnection
POI	Point of interest

(Continued)

Table 6.1 (Continued)

Acronym	Definition
PoTi	Position and time
PRS	Positioning reference signal
ReLU	Rectified linear unit
R-ITS	Roadside intelligent transport system
RNN	Recurrent neural network
RSSI	Received signal strength indicator
RTK	Real-time kinematics
SAE	Society of automotive engineers
SIR	Susceptible-infectious-removed
SoTA	State-of-the-art
SSR	State space representation
TS	Technical Specification
UE	User equipment
URLLC	Ultra-reliable low-latency communications
VAM	Vulnerable Road User Awareness Message
V2I	Vehicle-to-infrastructure
V2P	Vehicle-to-pedestrian
V2V	Vehicle-to-vehicle
V2X	Vehicle-to-Everything
VRU	Vulnerable road user

later not rely on the camera. The machine learning would bring the advantage of re-using a limited number of advanced people sensing devices, such as only a single device. For instance, a stereoscopic camera [6] can be used to provide high accuracy crowd estimation for a given environment. Furthermore, replacing the camera usage with machine learning would enable avoiding privacy invasion and significantly lower energy consumption as cameras would not need to be deployed vastly in many environments, and they would not need to always remain activated.

As an example people counting service, the CountMeIn [2] crowd mobility analytics service considers long-term high accuracy crowd estimation through wireless sensors (Wi-Fi scanning) of the environment and stereoscopic cameras as the auxiliary sensors for people counting. Anonymized (hashed and salted) Wi-Fi RSSI data are collected from any given environment, such that only RSSI levels and a unique hashed ID are considered to differentiate one device from another for short time periods (e.g. every five minutes). This way, it is not possible to understand any device belonging to a certain individual or track any device for any extended time period. In other words, individual tracking information or any personal data (e.g. data that could map to an individual) are not collected, whereas only signal strengths from the environment and timestamp information is available.

In any real scenario, a person could carry multiple Wi-Fi enabled devices such as wearables or no Wi-Fi enabled devices. Furthermore, static devices that do not belong to mobile entities (vehicles or people) in the environment can broadcast Wi-Fi probe packages. These packages could be anonymized as well, without any filtering or reading device information. However, the non-filtered data cannot be directly used for counting devices or individuals in a given environment. On the other hand, auxiliary sensors with high accuracies such as the advanced people counting sensor [6] provide people count-in and count-out events that could be leveraged to estimate the number of people that pass a certain area (e.g. narrow street or gate). Although the data from the advanced sensor can be leveraged, these sensors are costly, energy hungry, and privacy invasive in certain scenarios. Thus, CountMeIn considers leveraging the advanced sensor for a short training time and later replace it during its long-term operation with only wireless measurements, whereas still maintaining the accuracy through the trained machine learning model. Here, the machine learning models can learn the correlations between Wi-Fi and stereoscopic camera measurements.

Figure 6.2 illustrates the analytics pipeline of the CountMeIn service. The pipeline includes preprocessing, machine learning model building, machine learning inference steps. The results of the machine learning (CountMeIn outputs) are shared with an external system, as well as the CountMeIn dashboard, for deployment and visualization of insights. For the machine learning models, the selection includes various polynomial regression models with degrees ranging from 3 to 12, as well as neural network architectures. The machine learning model is trained for a minimum one week and for two weeks, time period for

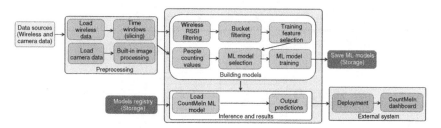

Figure 6.2 Analytics pipeline of CountMeIn. CountMeIn consists of wireless scanners and auxiliary sensor for people counting.

the optimal performance. The machine learning models with the initial filtering achieve error reductions for minute-level and hourly crowd estimation compared to state-of-the-art models. The error reductions are up to 44% for minute-level and 72% for hourly crowd estimations [2].

As the second example for people-centric data analytics, let us discuss the group inference using wireless data. We consider Group-In [1] as the analytics service for understanding existence of the groups in crowds, the number of people groups, and the group sizes.

Leveraging a single wireless scanner for understanding groups is considered in the literature. However, the RSSI data from wireless devices are highly volatile and noisy. Depending on the environment characteristics of deployment location of the scanner, the values may differ highly. For instance, even in the case of no obstacles or environmental effect, a scanner can provide inputs from two wireless devices that are at the same distance from the scanner, whereas in different directions. To provide high accuracy group inference and tackle the above-mentioned boundaries, Group-In considers leveraging inputs from multiple wireless scanners in the environment where each mobile device can broadcast wireless data (i.e. Bluetooth advertisement packets) and the data can be captured by multiple scanners. This data from multiple scanners can be combined to create a multi-modal wireless data.

Figure 6.3 illustrates the analytics pipeline including the pre-processing, building machine learning models, and prediction (group inference). The output predictions of Group-In are later leveraged by an external system for deployment and visualization through the Group-In dashboard. In the machine learning model, the data coming from different mobile devices through multiple wireless scanners are combined and aggregated for the given time interval, such as 60 or 120 seconds. Two different algorithms are considered for the combination and aggregation, one considering a centralized server and the other for a decentralized setup where each device can do initial aggregation and combination. Using the combined and aggregated data, the mobile devices are represented as a social network through a graph

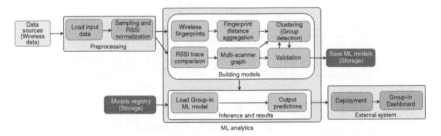

Figure 6.3 Analytics pipeline of Group-In. Group-In leverages multiple scanners' correlations with each other to extract people groups from crowds.

model. In the graph model, each node represents a mobile device, and each edge represents the closeness between a pair of mobile devices. Later, graph clustering algorithms such as DenGraph [7] (a modified version of DBSCAN [8] for graph data), HCS [9], and MaxClique [10] are applied for inferring community structures on the graph model.

6.1.1.2 Example Experimental Results from Crowd Mobility Analytics: Group Inference

The Group-In service is first implemented as a prototype, and it is initially tested through controlled data collection campaigns. As described in detail in [1], Bluetooth beacon devices (which broadcast Bluetooth advertisement packages regularly) are considered as the mobile entities and Raspberry PI devices are considered as the wireless scanners. The data collection is conducted in indoor setup with various settings, such as beacons distributed in different rooms, in the same room and with different distances. Moreover, different static and mobile behaviors are tested (e.g. random movement vs. straight movement). Each controlled experiment is marked with ground-truth for performance analysis.

As reported in detail in [1], the Group-In tackles various scenarios for accurate group inference and in most of the controlled experiments provides more than 80% accuracy in terms of pairwise group estimates (if two entities belong to same group or not) or Jaccard metric (the ratio for the intersection of actual groupings and observed groups by Group-In). The initial observations provide confidence that the system can provide high accuracy group inference for static and mobile scenarios.

For outdoor scenarios, similar controlled experiments should be conducted to test the applicability. Due to limitations of ground-truth collection in large urban environments regarding people groups, controlled experiments would help understand the challenges due to new elements in outdoor setups such as vehicles. Moreover, as the number of people and devices would vastly increase, the proposed centralized/decentralized algorithms and graph models might create large computational overhead. Therefore, before deploying in a large urban environment, the system should be tested for its scalability.

As the first step, the initial tests were conducted for the computational scalability of the approach with anonymized smart city data with, in most cases, more than 100 anonymized devices (up to 250). In this scenario, the computation in the analytics pipeline needs to aggregate, combine data from all devices, and apply clustering afterward. The setup of the data collection is described in detail in [1]. Figure 6.4 include observations of Group-In applied into the smart city dataset for one-week period ($\Delta t = 30$ sec sampling time and $T = 10$ min time interval). Here, wireless fingerprint matching of Group-In and HCS with pretrained (fixed) parameter values is used. The relation between group size and the number of

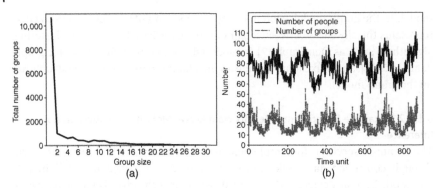

Figure 6.4 (a) Distribution of groups by group size observed, (b) (from [4]) Number of people vs. groups.

groups is given in Figure 6.4b. The number of groups sharply decreases with the increase in group size. Moreover, the relation between the number of groups and the number of detected people is shown in Figure 6.4. In this figure, each time unit corresponds to a time interval ($T = 10$ min). There exists a positive correlation between the number of groups and the number of detected people, and the data has a daily trend with a peak value (up to 110 people) every day. These observations do not provide much real insight on crowdedness or the accuracy of Group-In in the urban environment due to lack of ground-truth in the dataset. However, the system could be functionally applied to a significantly larger dataset and be able to produce results in simulations of real-time behaviors.

Table 6.2 shows the system scalability results in terms of the computing time, considering different time intervals from 2 to 60 minutes of smaller-size cluster of sensors (Cluster 2) and larger-size Cluster 1. The clusters represent two areas in

Table 6.2 Computation times considering different time intervals from 2 to 60 min of smaller-size Cluster 2 and larger-size Cluster 1.

Time int. (min)	Cluster 1			Cluster 2		
	# people	DB query time (sec)	Group inference time (sec)	# people	DB query time (sec)	Group inference time (sec)
2	43.1 ± 7.19	0.55 ± 0.18	0.75 ± 1.12	2.21 ± 0.59	0.05 ± 0.02	0.06 ± 0.03
5	59.4 ± 9.39	1.28 ± 0.38	1.85 ± 2.11	2.61 ± 1.05	0.10 ± 0.04	0.12 ± 0.04
10	80.9 ± 14.1	2.49 ± 0.67	3.97 ± 3.94	3.85 ± 1.95	0.18 ± 0.06	0.21 ± 0.07
30	153 ± 37.2	7.35 ± 1.74	20.2 ± 14.6	12.5 ± 5.13	0.46 ± 0.15	0.65 ± 0.21
60	232 ± 72.8	15.5 ± 3.34	79.4 ± 64.2	19.7 ± 9.36	0.86 ± 0.28	1.42 ± 0.52

the cities, where Cluster 1 represents a more crowded area with a higher number of wireless scanners. The number of people detected at each time interval, DB query time, and the total group detection time (which includes both query and computation) are listed in this table. For instance, for $T = 10$ min, the time it takes to create the group detection outputs is only a few seconds (for Cluster 1 having around 80 people). Therefore, the system is able to operate as a stream-based service for city-wide group monitoring. The results show the applicability of Group-In in the future to even larger cities and support real-time group detection. Group-In can analyze the data for long time periods and creates various group monitoring statistics.

6.1.2 Flow Monitoring

6.1.2.1 Introduction and Related Work

User equipment (UE) location data can provide varying types of information to be exploited both for network management actions and third party applications. A major category of these analytics and artificial intelligence (AI)/machine learning (ML) methods refers to the exploitation in various contexts (e.g. indoor, outdoor, mixed, vehicular, and public transportation-related) of attributes such as location, dwell time, orientation, speed, and trajectories per UE in order to derive insights related to people mobility aspects in an area, i.e. identifying point of interests (POIs) where a large number of people tend to gather, determining usual paths/trajectories people or vehicles/public transport vehicles follow, or even predict the trajectory of a specific UE.

The latter is a popular subject that has varying applications, such as network optimization, crowd management, and people flow monitoring for security and safety purposes, venue/mall/airport management, and smart targeted retail, traffic (pedestrian and vehicle) management for the transportation sector, and self-driving vehicles. In more detail, relevant research activity on trajectory prediction has been ongoing utilizing ever-increasing sequential online and offline data, such as video and audio. More baseline prediction approaches, such as the Kalman filter, linear, or Gaussian regression models, have been however surpassed with the advancement of /acDL due to the complexity of human behavior patterns and non-linearity of trajectories [11, 12]. While the purpose of this section is not to describe the mathematical formulations and various approaches in detail, some introductory descriptions follow.

The main deep learning (DL) approaches, and specifically Recurrent neural network (RNN)-based architectures, have been designed to support the processing of sequences of values [13]. In fact, RNNs can learn by being trained to predict the next symbol in a sequence, with the most common implementations being long short-term memory (LSTM) [14] and gated recurrent units (GRUs) [15] with the latter being more recent and simpler implementation-wise. In the same context, RNN-based autoencoders learn to encode a variable-length sequence into a latent representation of fixed-length and then decode this representation back into a sequence [16].

All these approaches – in the context of trajectory prediction – are based on the sequence of position coordinates regardless of the presence of other humans, vehicles, etc.; however in [17] authors introduce the idea of the social context in the trajectory prediction, i.e. modeling interactions between mobility users (pedestrians, vehicle drivers, passengers, etc.) sharing the same scene. Social LSTMs in [18] allowed for "social pooling", thus modeling the social interaction of all road users in a specific radius, whereas social generative adversarial networks s (GANs) [19] introduced a different version of the pooling process, considering all users of the scene. In addition to the aforementioned, Transformer models, presented in 2017 [20], were applied for trajectory prediction [21], while their traditional architecture was further enhanced by capturing spatial–temporal dependencies between agents [22].

In this context, authors in [23] compared the most recent and state-of-the-art (SoTA) approaches, i.e. Social GAN and Trajectory Transformer network with a proposed GRU implementation including recurrent units and attention mechanisms. These approaches and the results yielded will be further described in Section 6.1.2.2.

6.1.2.2 Selected DL Approaches and Results for Trajectory Prediction

Social GAN GANs are a category of deep approaches where two components are trained as adversaries to maximize effectiveness. In particular, they consist of two components – the generator and the discriminator. The first tries to understand the underlying characteristics of an input dataset to synthesize similar data as close as possible to the real-world ones, while the latter aims at differentiating between real and synthesized data [24]. The Social GAN architecture also includes a pooling module, the purpose of which is to quantify the social interactions of the different agents. More specifically, the architecture can be summarized as follows:

- Input coordinates as well as relative positions of all other UEs/mobility agents with respect to the input user are fed to the pooling module, and specifically a single multi-layer perception (MLP) followed by the symmetric Max-Pooling operation. This approach differs from social pooling, where only users in a certain radius are taken into consideration.
- The generator produces a sequence based on a sequence input. It consists of an LSTM-encoder, a pooling module, and an LSTM-decoder, so that the encoded representation of the input is subjected to a pooling operation and is then fed to the decoder, thus incorporating social information in the decoder input.

- The discriminator focuses on the classification of a sequence as real or fake and consists of a LSTM-Encoder and a dense layer with softmax activation function, producing the more "acceptable" paths.
- Binary cross-entropy function is used for the discrimination loss, whereas L2-norm is used as the generator loss.

Transformer Model A transformer model predicts the future trajectory of a UE based on its previous positions and employs an encoder–decoder approach. However, it does not incorporate recurrence or considers social context, as in Social GAN and employs attention mechanisms [20]. In more detail:

- The timestamp t is added as an extra dimension in the input sequence vector of the encoder.
- For the encoding and decoding components, a feed-forward fully connected module, called the Attention module, is employed.
- Attention [20] is modeled as follows: For each sequence input, the dot product is computed with every other sequence, scaled with the dimensionality of the two sequences and a softmax function is applied to weight the various sequence entries. The process is feed-forward and thus highly parallelizable, as opposed to the previous method of GANs.

GRU Model with Attention Presented in 2014 [25] GRU is an RNN-based method which – compared with the LSTM – combines the forget and input gate in an update gate. Authors in [26] suggest that GRU architectures outperform LSTMs and are easier to configure as the training parameters of each unit are reduced, making the whole architecture more resilient to overfitting. As an added advantage, they are also more computationally efficient.

In [23], a GRU approach is used for trajectory prediction. To further improve model performance, an attention mechanism in the encoding block of the model, imitating the transformer approach [21] is chosen, using a scaled dot product attention where the scalar parameter is now a learnable weight. As also shown in Figure 6.5:

- The decoder is fed the result of the attention module, consisting of a GRU layer and a dense layer with ReLU as an activation function.
- The output linear layer of the network produces the output
- The loss function chosen was mean squared error (MSE).

Datasets The models were evaluated on both simulation and open-source datasets that have been generated under controlled mobility situations. Specifically, a mixed pedestrian/vehicular mobility simulation was done using a commercial

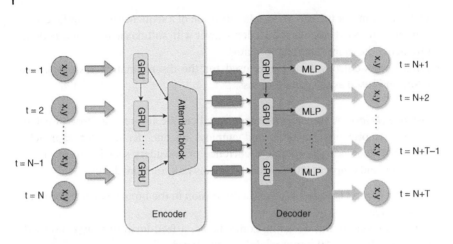

Figure 6.5 GRU with attention architecture.

simulator *IncelliSIM* to emulate the mobility patterns that occur in mixed indoor–outdoor conditions. In the same context, the ETH/UCY dataset. ETH is a pedestrian dataset, containing images [27] and UCY is a pedestrian dataset with rich multi-human interaction scenarios [28].

Results Quantitative results were compared in terms of final displacement error (FDE), i.e. the average for all trajectories Euclidean error of the final data point predicted, average displacement error (ADE) – the MSE for all data points predicted averaged for all trajectories and their mean value mean distance error (MDE). Indeed, for all types of metrics, the GRU-based approach outperformed the other methods. Furthermore, in the case of MDE, it was improved by 5.7% in the ETH/UCY dataset and 29% in the simulated dataset. Table 6.3 shows the MDE for both ETH/UCY and IncelliSim dataset.

Furthermore, from visual inspection it was evident that both Transformer and GRU models can produce a realistic outcome, despite not including social

Table 6.3 Displacement error for ETH/UCY and IncelliSIM datasets [23].

Model	ETH/UCY dataset			IncelliSim dataset		
	FDE	ADE	MDE	FDE	ADE	MDE
Social GAN	1.24	0.60	0.92	6.96	3.54	5.25
Transformer	0.95	0.43	0.69	1.5	0.62	1.06
GRU	**0.88**	**0.41**	**0.64**	**0.92**	**0.57**	**0.75**

Bold values denote the lowest errors when comparing the three methods.

dependencies. On the other hand, they failed to capture sudden and/or unusual behaviors, such as sudden acceleration or direction shift. To this end, future work should involve exploring additional information as input, e.g. angle/direction and velocity, enriching the input vector space.

6.1.3 COVID−19 Contact Tracing

6.1.3.1 Introduction and Related Work

Mobility between geographical units was one of the main magnitudes monitored during the COVID-19 pandemic, since it could offer insights into the propagation of the virus, the mobility restrictions that could be effective to stop it, etc. Mobile networks have intrinsically the ability to monitor mass mobility, showing which geographical units are mostly interconnected. To do this, two elements are required: a scheme for dividing the land (for instance, provinces, or Sub-City Districts), and a technology to establish the origin and destination divisions of a user. The aggregation of movements between the different divisions in a certain interval of time gives an idea of their interconnectedness. Figure 6.6 shows how these measures can be taken using a 5G network. The location precision required is usually low, so even techniques such as trilateration with macro-cells are enough for the estimation. To preserve user privacy, only flows that are higher than a certain number of individual users can be reported; ignoring cases where only a few users go from one area to another in a specific interval of time.

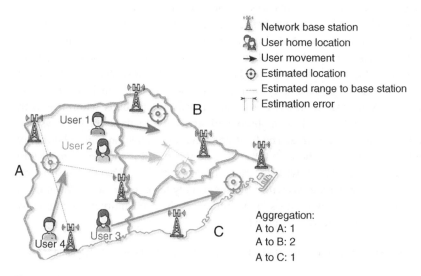

Figure 6.6 Measurement of mobility among geographical units.

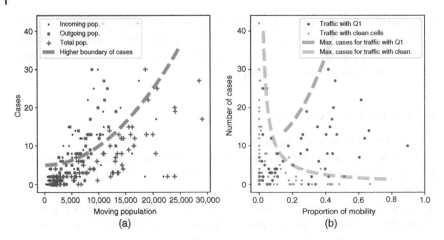

Figure 6.7 Number of cases vs. mobility (a) and traffic profiles of all the cases (b).

Relating the mobility to the COVID-19 spread is then trivial. For each area, a tuple of the total mobility (incoming or outgoing users) with the number of cases can be defined. The results of this matching are shown in Figure 6.7a for the province of Málaga (Spain). Each point represents a cell with a population of at least 5000.[1] It can be seen that a low mobility tends to produce a low rate of infections, conforming a boundary of the maximum number of cases based on mobility, shown with a dashed gray line. On the other hand, a high mobility increases the higher maximum number of possible cases, although other factors seem to actually decide whether the number of cases will or will not be high.

This mobility study can be further refined, determining not only the total mobility of a geographical area, but also with which other types of areas (in terms of COVID-19 propagation) does it relate. Five categories can be defined: Areas with no cases, and quartiles from Q1 to Q4. Q1 represents those that have the most cases, whereas Q4 represents those that have more than 0, but it has the least cases. Figure 6.7b shows the two extremes, that is, the mobility with areas with zero cases (gray dots) and the mobility with areas of Q1 (dark gray dots) vs. the number of cases in the area. The light gray line represents the higher boundary of infections for the traffic with Q1 areas, and the gray line the higher boundary given the traffic with areas with no cases. These results show that areas with a high traffic proportion with areas with no cases usually have fewer cases. Although a high traffic exchange with areas in Q1 influences the probability of having many cases, they are not the determining factor.

---

1 Instituto Nacional de Estadística (INE). Estudios de Movilidad a Partir de la Telefonía Móvil.

In [29] it was discussed how 5G and beyond can play a primary role in contact tracing and group movement monitoring. In particular, contact tracing based on 5G location-based analytics benefits from the pervasive deployment of cellular networks, the several years of effort to design cellular standards for localization and analytics, and the best practices of cellular operators to handle location data.

With the COVID-19 spreading, several restrictions on the human mobility (a dominant factor that drives the evolution of the disease) have been enforced by national authorities to slow down the contagion. It is therefore necessary to predict the evolution of the pandemic to support national authorities in planning timely and specific countermeasures. Furthermore, the predictions can be used by medical care systems to plan in advance resource allocation (for example, to face up a predicted high occupation of the intensive care units).

6.1.3.2 Selected Approach and Example Results from Contact Tracing

A stochastic model has been established to predict the spatiotemporal evolution of COVID-19 disease. The model is composed of two interacting sub-models: (i) a compartmental epidemiological model; and (ii) a stochastic human mobility model. The former has been developed as an extension of the classical SIR model [30] and aims at describing the time-varying number of people belonging to specific compartments. Each compartment represents the epidemiological situation of the people belonging to it. The human mobility model probabilistically describes the time-varying people flows among different interacting spatial locations. The whole model is mathematically described by a system of stochastic differential equations whose solution provides the prediction of the number of people belonging to the compartments. In particular, six different compartments have been identified, namely susceptible, infectious, undetected infectious, hospitalized, quarantined, and removed. The probabilities associated with the people flows have been inferred by using the mobility dataset provided by [31].

Case studies are presented for two scenarios in which the evolution of infections and hospitalizations is predicted for Emilia-Romagna Italian region considering the first 30 days of the outbreak period. Figure 6.8 shows the comparison between predicted and real infections, and predicted and real hospitalizations, respectively. For each prediction, upper (95%) and lower (5%) confidence bounds are represented. The real data are provided by the Dipartimento della Protezione Civile, Italy. The obtained results prove the high prediction accuracy of the developed model and the innate connection between human mobility and the spreading of the COVID-19 disease.

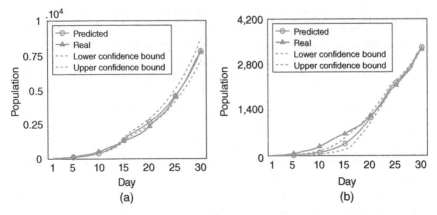

Figure 6.8 Comparison between: (a) predicted and real infections; and (b) predicted and real hospitalizations.

6.2 Localization in Road Safety Applications

The localization and analytics in vehicular systems have the potential to significantly improve human mobility and quality of life; hence, this is the main focus of the section. The evolution of the vehicular systems is moving toward ever more connected and fully automated vehicles. Such high level of autonomy leverages two main enablers, among others: local-awareness based on accurate positioning and sensing, and ultra-low latency communications among vehicles within a shared network infrastructure. These functionalities allow vehicles to develop a shared perception of their surroundings and make decisions based on local views and expected maneuvers from nearby users. The combination of ultra-low latency communication with accurate positioning and sensing leads the way toward a safer transportation system with the goal of achieving zero road deaths and a better traffic flow.

The enhancement of V2X technology in 5G, which allows any vehicle to interact with any other road element (i.e. roadside units, pedestrians, network and infrastructures), enables URLLC with high data rates [32, 33]. Given such unprecedented combination of URLLC and high localization accuracy, 5G is the first technology that has the potential to meet the very stringent requirements of road safety applications. Nevertheless, several of the use cases presented by the industrial associations refer to extremely stringent latency and accuracy requirements, which might not be met by the 5G technology alone, especially in challenging operating conditions. In these scenarios, it is therefore necessary to employ advanced localization and sensing techniques, to hybridize with non-radio technologies, while remaining in accordance with the 5G architecture [34–37].

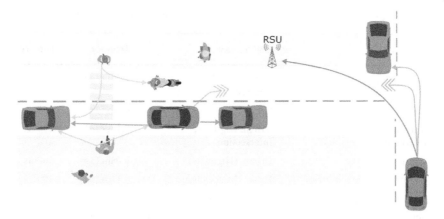

Figure 6.9 Illustration of four use cases with stringent positioning requirements in an urban scenario: coordinated driving maneuver, interactive VRU crossing, infrastructure assisted environment perception, and left turn in multilane street with oncoming traffic.

6.2.1 Safety-Critical Use Cases and 5G Position-Related Requirements

6.2.1.1 Introduction and Related Work

Positioning, including relative positioning, accurate inter-vehicle ranging, and radar capabilities are the key functionalities for a large number of safety-critical services, which operate under small inter-vehicle distances.

The standardization of 5G-V2X is guided by advanced use cases that extend the basic applications defined in LTE-V2X. The 3GPP and the 5GAA are the two entities that have defined vehicular scenarios in 5G, by categorizing the use cases and indicating the key performance metrics. Figure 6.9 illustrates multiple safety-critical use cases in a typical urban scenario. In particular, Table 6.4 summarizes the performance requirements of a few selected use cases with stringent requirements in terms of accuracy, latency, environment of use, and reliability for the use cases described in the following:

- Coordinated, cooperative driving maneuver: A vehicle shares its intention to perform a certain action (e.g. lane change) with other traffic participants potentially involved in the maneuver, which interact with the vehicle to confirm or decline the planned maneuver.
- Interactive VRU crossing: A VRU, such as a pedestrian or cyclist, shares his or her intention to cross a road and interacts with vehicles approaching the area in order to improve safety for VRUs and awareness for vehicles.
- Infrastructure assisted environment perception: A vehicle enters a section of the road that is covered by infrastructure sensors, e.g. roadside units (RSUs)

Table 6.4 Example of safety-critical use cases and service level requirements.

Use cases	Environment	Accuracy	Latency
Coordinated, Cooperative Driving maneuver	Urban, Rural, Highway, Intersection	1.5m (3σ)	160ms
Interactive VRU Crossing	Urban	0.2m (3σ)	100ms
Infrastructure Assisted Environment Perception	Urban, Highway, Intersection	0.1m (3σ)	100ms
Drifting out of lane	Urban, (Highway)	0.08m (1σ)	200ms
Left turn in multilane street	Urban, Intersection	0.13m (1σ)	10ms

The reliability required for all these use cases is 99.9%. According to the preliminary performance assessment from 3GPP [39], the accuracy requirements can generally not be achieved by the 3GPP Rel. 16, with the only exception of the first use case. Similarly, the latency requirements can be achieved only for the first and fourth use cases.

and subscribes to receive the information from the infrastructure containing environment data from dynamic and static objects on the road.

We also find of interest two additional use cases defined by the Euro NCAP Test Protocol for road traffic collision avoidance:

- Drifting out of lane: A sensing system detects a drift out of the lane and warns the driver or corrects the driving path autonomously before the ego car exits the lane.
- Left turn in multilane street with oncoming traffic: A vehicle performs a left turn across the path of the oncoming traffic.

6.2.1.2 Example Results for Safety-Critical Use Cases

The 5G positioning service levels 4 and 6 defined in [38] for Release 17 are currently the closest to the requirements defined in Table 6.4 as they refer to 0.3 m of horizontal accuracy and latency of 15 and 10ms, respectively. While the accuracy requirements in Table 6.4 refer to absolute positioning, there are many use cases where relative positioning is more relevant (e.g. platooning and group start services). Therefore, relative positioning, accurate inter-vehicle ranging, and radar capabilities become the key functionalities in these safety-critical services, which operate under small inter-vehicle distances.

Furthermore, several use cases rely on positioning, and in particular the positioning of the vehicle itself, but the VRU and environment perception use cases also include sensing aspects, where the shape and type of objects may need to be determined through passive (radar-like) measurements. Ultimately, vehicular safety applications necessitate extreme positioning accuracy and latency. In real operating conditions, 5G technology alone is not always capable to meet these requirements, as visualized with colors in Table 6.4. Indeed, preliminary studies conducted within the 3GPP [39] show that the horizontal positioning accuracy can achieve sub-meter levels in 90% of cases only in ideal conditions and that physical layer latency often exceeds 100ms. Hence, progress is needed in terms of enhancing the existing 5G standard and in terms of new enabling technologies beyond 5G.

6.2.2 Upper Layers Architecture in ETSI ITS Standard

6.2.2.1 Introduction and Related Work

C-ITS protocol stack and services: European Telecommunication Standards Institute (ETSI) ITS Technical Committee established in 2007 is responsible for the cooperative intelligent transport systems (C-ITS) standards development, interoperability, and conformance. European focused standardization activity is widely recognized worldwide, and ETSI ITS specifications [40] define four ITS station (ITS-S) sub-systems (personal, vehicle, roadside, and central) with ITS stations which communicate in ITS environment. ITS communications architecture is defined based on Open System Interconnection (OSI) model extended for ITS (see [41]). The higher layers are intended to be access technology-agnostic to allow forward compatibility. The protocol stack is defined for basic ITS-S functionality (host) and includes the following layers:

- Applications layer: – Support for road safety, traffic efficiency, infotainment, and other use cases.
- Facilities layer: – Applications, information services, and session support, e.g. Global positioning system (GPS) positioning, state monitoring (car engine, lights), messages, and time management.
- Networking and transport layer: – Protocols for data delivery and routing between ITS-S stations, e.g. internet protocol version 6 (IPv6) support, handover between access technologies.
- Access layer: – Internal and external ITS-S communication using various media (ITS-G5, wireless fidelity (WIFI), Ethernet, Cellular, e.g. LTE-V2X, Bluetooth, GPS).
- Management entity: – Configuration of ITS-S station and cross-layer information exchange.
- Security entity: – Privacy and security services.

Furthermore, the following ITS services are defined in the first two releases of the ETSI ITS standard, leveraging ITS-S positioning and sensing capabilities to support road traffic safety applications:

- Cooperative awareness basic service: [42] – Facilities layer service to allow road users and infrastructure be informed about object state (e.g. vehicle time, position, motion state, activated systems) and attributes (e.g. dimension, type, role). Information is exchanged using periodic cooperative awareness message (CAM) using vehicle-to-vehicle (V2V) or vehicle-to-infrastructure (V2I) with predefined syntax, semantics, and handling.

- Decentralized environmental notification: [43] – Facilities layer service triggered by Road Hazard Warning applications includes decentralized environmental notification (DEN) messages containing road hazard (e.g. icy road) or abnormal traffic conditions (type, position, duration). Notification dissemination (V2V or V2I) uses pre-defined geographic area of the receiving ITS stations close to the detected event. Information is presented to the driver to react or forwarded to central ITS for traffic management purposes. Protocol (including messages syntax and semantics) is designed to manage situation when detected event persists after originating ITS station is far from its location.
- Local dynamic map: [44] – Database of time and location referenced objects influencing or influenced by road traffic. Digital map data store maintains useful moving or stationary objects info, e.g. lane specific info, pedestrian walking, bicycle paths, traffic lights, objects sensed by other users. Facilities layer function describes object dependencies, relationships, and timestamps. CAM and DEN messages are data sources for local dynamic map (LDM). Standard describes LDM functional behavior, functions, interfaces, and data objects for safety applications.
- Collective perception service: [45] – Enables ITS-Ss information sharing of objects (other users, obstacles) detected by local perception sensors (cameras, radars, etc.) increasing situational awareness. It defines Collective Perception Messages syntax and semantics which provide information about the disseminating ITS-S, its sensors and detected objects reducing environment uncertainty in a cooperative manner.
- Vulnerable road users awareness basic service: [46] – Based on the vehicle-to-pedestrian (V2P) communication it enables the operation of vulnerable road user (VRU) awareness messages including time, position, motion state, and other attributes to enhance the protection of pedestrians, cyclists, motor cyclists, and animals posing potential safety risk. It also includes the support for profiling VRUs, VRU clustering, and dynamic service operation based on the vulnerability context.

Positioning function in C-ITS architecture: To be able to support road safety services, ITS-Ss needs to have precise position and common time information. Minimum accuracy, integrity, and reliability requirements are dependent on the supported C-ITS applications and may evolve, e.g. for vehicle platooning use case, position, and time needs to be exchanged more than twenty times a second. As a result, in Release 2, ETSI ITS standard defined position and time

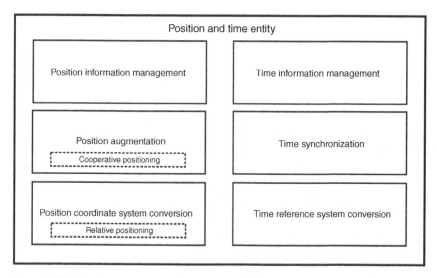

Figure 6.10 C-ITS position and time entity functional architecture (adapted from [47]).

(PoTi) services (as part of the facilities layer) which specify related functional and operational requirements in ITS applications. In particular [47] describes functional architecture, exchanged messages, and protocol to facilitate improved position and time accuracy in ITS-Ss. PoTi entity is an essential part of the C-ITS system as it manages the position (including information management, augmentation and coordinate system conversion functions) and also time (information management, synchronization, and reference system conversion functions) used by other ITS-S layers (see functional architecture in Figure 6.10). It leverages information from GNSS and other sensors (inertial, odometer, cameras, radio ranging, map matching, etc.) to manage kinematic and attitude state of the ITS-S which includes time, position, velocity, acceleration, orientation, angular velocity, and other motion information with supporting confidence information of these variables. In addition, ITS-S may use various augmentation methods to improve the time and location accuracy. They are based on the external assistance data or measurements helping to improve positioning performance and eliminate system errors. Satellite-based augmentation may be based, e.g. on European geostationary navigation overlay service (EGNOS), Real-Time Kinematics, or Sirius Satellite Radio service. Augmentation based on the cooperative positioning between ITS-Ss may leverage EGNOS, Radio Technical Commission for Maritime service or state space representation (SSR) correction information. PoTi standard recognizes GNSS Positioning Correction, Road-ITS-Station, and cellular-based augmentation services improving the position and time estimates.

Support for VRU clustering: At a large urban intersection, where there is often a high density of road users with dynamic movement, ensuring user safety is a critical requirement especially for the VRUs. To successfully assess a road safety risk for each VRU in ETSI C-ITS architecture, there is a periodic exchange of standard VRU messages required between different VRU devices equipped with personal ITS-S and a roadside ITS-S (R-ITS) or vehicle ITS-S at the crossing.

In such crowded scenarios, this may require a prohibitive amount of communication and processing resources, e.g. in dedicated 5.9 GHz ITS band since separate V2P broadcast messaging sessions are to be initiated for each VRU device in a safety risk zone causing a network congestion and a message loss. However, with a VRU clustering solution, VRU devices can be virtually grouped together and only the cluster head VRU (leader) will continuously exchange messages with the R-ITS (which may control cluster formation and maintenance), requiring a single communication session for the cluster. The cluster parameters are communicated using the VAM as described in Section 6.2.2.

VRU clusters can be supported by an application using localization and group inference analytics as described in Section 6.1.1. Several criteria need to be satisfied for the formation and maintaining of VRU clusters. These include the relative speed and relative distance between members of a cluster as well as the number of VRUs already existing in a cluster. Detailed operational requirements for VRU clusters are defined in ETSI TS 103 300-3 standard [46].

The Shibuya intersection in Tokyo, Japan, famous for being the world's busiest intersection, is a good VRU clustering study scenario. At its busiest, as many as 3000 people cross in different directions [48] ; thus, the performance of VRU clustering is evaluated using a simulation model of the intersection. Here, a wide range of VRU velocities is considered which represents both slow moving pedestrians (e.g. children and the elderly) and fast moving pedestrians (e.g. runners). The maximum speed difference inside a cluster is kept at 5% according to the ETSI standard [46], and the VRUs are tracked until they are out of the safety risk zone, i.e. they have safely crossed the intersection. The VAM messaging overhead is calculated by the sum of the total number of exchanged messages in the VRU system, which is an ensemble of ITS stations interacting with each other to support VRU crossing, e.g. VRU device ITS-S, vehicle ITS-S, R-ITS, and C-ITS. The frequency of a standard VAM message is 2000 ms [42].

6.2.2.2 Example Results for ITS

Figure 6.11 shows the cluster size capacity when VRUs are grouped according to their relative speeds. It can be seen that there is an inverse relationship between the cluster size and VAM messaging overhead since a higher constraint results in a fewer total of VRU cluster heads, and thus a decrease in the messaging overhead. More interestingly, however, we observe a plateauing of the VAM messaging traffic

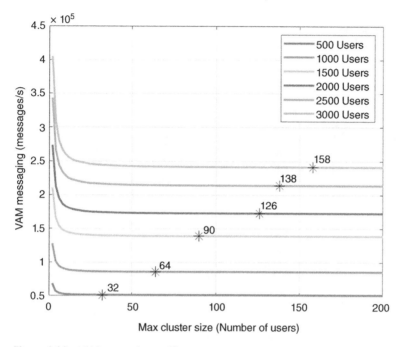

Figure 6.11 VAM messaging traffic.

after a certain cluster size for different VRU groups as indicated by the starred points. These results are generic and indicate that cluster sizes up to the maximum of the indicated starred values would provide some communication benefit.

On the other hand, Figure 6.12 illustrates the actual statistics of formed clusters in the Shibuya intersection model. It can be seen that the numbers and sizes of VRU groups increase with increasing VRU numbers. Specifically, at the busiest periods, with 3000 users crossing the intersection, 81 clusters each made of 36 VRUs are formed while at less busy periods, with 500 users, 54 clusters of 9 VRUs each are formed. Note that although 3000 is the current maximum number of users at the Shibuya intersection, some extra capacity would need to be considered when designing the commercial system.

Figure 6.13 illustrates the gains of the clustering solution by comparing the messaging traffic with and without the clustering. It can be observed that VRU clustering achieves a lower VAM messaging traffic for the same conditions. In particular, the clustering solution provides reduction in the traffic and corresponding communication resources and spectrum usage by approximately 47% for lower VRU numbers and 60% for higher VRU numbers. These results are important, indicating that VRU clustering solutions are especially beneficial during the busiest traffic periods at larger intersections.

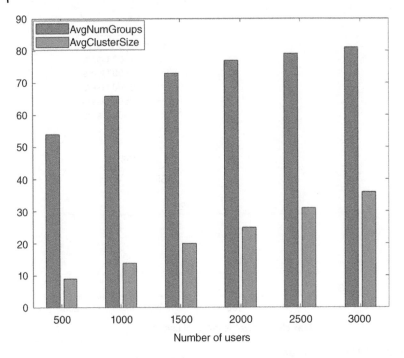

Figure 6.12 VRU clustering statistics.

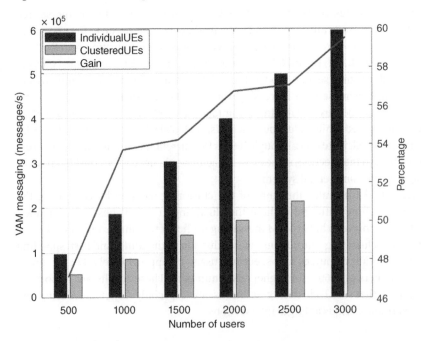

Figure 6.13 VRU clustering impact.

6.2.3 5G Automotive Association (5GAA) Activities

5GAA is a global cross-industry organization formed in 2016 with the intention to address connected mobility and road safety needs, promote and develop cellular V2X (C-V2X), and also support standardization and commercial adoption. It has rapidly expanded since and today among its 130 members there are car manufacturers, suppliers, telecom operators, and mobile equipment vendors. Its work focuses on influencing use cases harmonization, business models, technology roadmap evolution, spectrum allocation, security, regulatory, certification, and approval processes, and so far it has published many whitepapers, position papers, and studies covering these areas [49]. To this end, positioning is identified as one of key aspects in the definition of C-V2X service-level requirements and is used in the organization template for describing use cases [50]. Furthermore, in its recent report focusing on high-accuracy positioning for C-V2X [51], the organization has analyzed the following localization aspects:

- Positioning requirements in different V2X use cases: These cover traffic safety, traffic efficiency, and information services and depending on the use case complexity (i.e. velocity, vehicle density, environment), the positioning accuracy may vary significantly, e.g. from 0.1 m (3σ) for the most demanding (e.g. Tele-Operated Driving or High Definition Sensor Sharing) up to 1.5 m for less demanding (e.g. Emergency Break Warning or Lane Changing Warning) or even more (e.g. 50 m (1σ) for the Software Update or HD Content Delivery use case). The document also investigates positioning importance and requirements for society of automotive engineers (SAE) L4/L5 automated driving scenario.
- Positioning challenges: – Ensuring stable continuous high-accuracy positioning in complex and obstructed environments (e.g. urban, tunnel) may require integrated multi-source data fusion combining the satellite and the cellular network positioning, inertial navigation, radar, camera etc. while making sure the positioning cost is acceptable.
- Positioning system architecture: – It may be based on the UE-based positioning where the positioning calculation is performed in the user terminal leveraging a fusion algorithm of GNSS, inertial measurements, sensors, HD map, cellular network data and consists of terminal, network, platform, and application functional blocks. Conversely, UE-assisted positioning architecture uses the same functional blocks and is network centralized. The information fusion based on the input from the terminal and roadside infrastructure is performed in the network and then provided to the UE. Finally, the document also describes variants of the 3GPP sidelink-based positioning system architecture using direct communication between two UEs without relaying their data via the network – it requires sidelink positioning configuration, Sidelink

Positioning Reference Signal transmission, reception, measurements, and position calculation.

- Positioning KPIs: – These are mostly adopted 3GPP definitions described in [52] and include position accuracy, availability, latency, time to first fix, update rate, system scalability, continuity, reliability, integrity, time to alert, authentication, security/privacy, institutional compliance, consistency, update rate, power consumption, and energy per fix.
- Technologies for high-accuracy vehicles positioning: – There are various mechanisms described with their performance characteristics and standardization status:
  - GNSS based on RTK differential system: – Correction data is broadcasted over a cellular network to achieve centimeter-level accuracy.
  - GNSS based on SSR service: – Error state profile is broadcasted to the UE via SSR allowing sub-decimeter positioning.
  - Visual positioning based on sensors and HD map: – Position is calculated based on HD map object information and sensor measurements combined with RTK enabling centimeter-level accuracy.
  - Cellular network-based location using the radio signals: – Tens of meters, accuracy in LTE, a meter in 5G.
  - Sidelink positioning: – Either the UE or the network can calculate the position based on the transmitted reference signal allowing sub-meter accuracy. Improved method leveraging vehicle dynamics is also presented.
  - 5G mmWave and sensor fusion positioning: – Sub-meter accuracy could be potentially achieved by fusing a position information based on the 5G positioning reference signal (PRS) with base station position, range, angle-of-departure (AoD) measurements, and Inertial Measurement Unit data.
  - Vehicular positioning improvements: – Techniques such as cooperative positioning among terminals, improved synchronization, or multi-panel distributed antenna positioning are finally highlighted.

References

1 G. Solmaz, J. Fürst, S. Aytaç, and F.-J. Wu. Group-In: Group inference from wireless traces of mobile devices. In *Proceedings of ACM/IEEE International Conference on Information Processing in Sensor Networks (IPSN)*, pages 157–168, Sydney, Australia, April 2020.

2 G. Solmaz, P. Baranwal, and F. Cirillo. CountMeIn: Adaptive crowd estimation with Wi-Fi in smart cities. In *IEEE PerCom* 2022, Pisa, Italy, March 2022.

3 G. Solmaz, F.-J. Wu, F. Cirillo, E. Kovacs, J. R. Santana, L. Sanchez, P. Sotres, and L. Munoz. Toward understanding crowd mobility in smart cities through the internet of things. *IEEE Communications Magazine*, 57(4):40–46, 2019.

4 S. Bartoletti, L. Chiaraviglio, S. Fortes, T. E. Kennouche, G. Solmaz, G. Bernini, D. Giustiniano, J. Widmer, R. Barco, G. Siracusano, A. Conti, and N. B. Melazzi. Location-based analytics in 5G and beyond. *IEEE Communications Magazine*, 59(7):38–43, 2021.

5 F.-J. Wu and G. Solmaz. CrowdEstimator: Approximating crowd sizes with multi-modal data for internet-of-things services. In *ACM MobiSys*, pages 337–349, 2018.

6 HELLA Aglaia People Sensing Technologies. Advanced People Sensor APS-180E. http://people-sensing.com/, 2017.

7 T. Falkowski, A. Barth, and M. Spiliopoulou. DenGraph: A density-based community detection algorithm. In *Proceedings of the IEEE/WIC International Conference on Web Intelligence and Intelligent Agent Technology*, pages 112–115, Fremont, CA, USA, November 2007.

8 M. Ester, H.-P. Kriegel, J. Sander, and X. Xu. A density-based algorithm for discovering clusters in large spatial databases with noise. In *Proceedings of the International Conference on Knowledge Discovery and Data Mining*, pages 226–231, Portland, OR, USA, August 1996.

9 E. Hartuv and R. Shamir. A clustering algorithm based on graph connectivity. *Information Processing Letters*, 76(4–6):175–181, 2000.

10 K. Makino and T. Uno. New algorithms for enumerating all maximal cliques. In *Scandinavian Workshop on Algorithm Theory*, pages 260–272, Humlebaek, Denmark, July 2004.

11 S. Becker, R. Hug, W. Hübner, and M. Arens. An evaluation of trajectory prediction approaches and notes on the trajnet benchmark, 2018. URL https://arxiv.org/abs/1805.07663.

12 H. Jiang, L. Chang, Q. Li, and D. Chen. Trajectory prediction of vehicles based on deep learning. *2019 4th International Conference on Intelligent Transportation Engineering (ICITE)*, pages 190–195, 2019.

13 I. Goodfellow, Y. Bengio, and A. Courville. *Deep Learning*. MIT Press, 2016. http://www.deeplearningbook.org.

14 S. Hochreiter and J. Schmidhuber. Long short-term memory. *Neural Computation*, 9(8):1735–1780, 1997.

15 J. Chung, C. Gulcehre, K. Cho, and Y. Bengio. Empirical evaluation of gated recurrent neural networks on sequence modeling, 2014. URL https://arxiv.org/abs/1412.3555.

16 K. Cho, B. van Merrienboer, C. Gulcehre, D. Bahdanau, F. Bougares, H. Schwenk, and Y. Bengio. Learning phrase representations using RNN

encoder-decoder for statistical machine translation, 2014. URL https://arxiv
.org/abs/1406.1078.

17 A. Robicquet, A. Sadeghian, A. Alahi, and S. Savarese. Learning social eti-
quette: Human trajectory understanding in crowded scenes. In B. Leibe,
J. Matas, N. Sebe, and M. Welling, editors, *Computer Vision – ECCV 2016*,
pages 549–565. Springer International Publishing, 2016.

18 A. Alahi, K. Goel, V. Ramanathan, A. Robicquet, L. Fei-Fei, and S. Savarese.
Social LSTM: Human trajectory prediction in crowded spaces. In *Proceedings
of IEEE Conference on Computer Vision and Pattern Recognition (CVPR)*, pages
961–971, 2016.

19 A. Gupta, J. Johnson, L. Fei-Fei, S. Savarese, and A. Alahi. Social GAN:
Socially acceptable trajectories with generative adversarial networks, 2018.
URL https://arxiv.org/abs/1803.10892.

20 A. Vaswani, N. Shazeer, N. Parmar, J. Uszkoreit, L. Jones, A. N. Gomez,
L. U. Kaiser, and I. Polosukhin. Attention is all you need. In I. Guyon,
U. V. Luxburg, S. Bengio, H. Wallach, R. Fergus, S. Vishwanathan, and
R. Garnett, editors, *Advances in Neural Information Processing Systems*,
Volume 30. Curran Associates, Inc., 2017.

21 F. Giuliari, I. Hasan, M. Cristani, and F. Galasso. Transformer networks for
trajectory forecasting. In *Proceedings of International Conference on Pattern
Recognition (ICPR)*, pages 10335–10342, Virtual-Milano, Italy, January 2021.

22 C. Yu, X. Ma, J. Ren, H. Zhao, and S. Yi. Spatio-temporal graph transformer
networks for pedestrian trajectory prediction, pages 507–523. Springer-Verlag,
Berlin, Heidelberg, 2020. ISBN 978-3-030-58609-6.

23 Y. Filippas, A. Margaris, and K. Tsagkaris. Deep learning approaches for
mobile trajectory prediction. In *Proceedings of the IEEE Globecom Workshops
(GC Wkshps)*, Madrid, Spain, 2021.

24 I. J. Goodfellow, J. Pouget-Abadie, M. Mirza, B. Xu, D. Warde-Farley, S. Ozair,
A. Courville, and Y. Bengio. Generative adversarial networks, 2014. URL
https://arxiv.org/abs/1406.2661.

25 K. Cho, B. van Merrienboer, D. Bahdanau, and Y. Bengio. On the properties of
neural machine translation: Encoder-decoder approaches, 2014. URL https://
arxiv.org/abs/1409.1259.

26 R. Jozefowicz, W. Zaremba, and I. Sutskever. An empirical exploration of
recurrent network architectures. *Journal of Machine Learning Research*,
37:2342–2350, 2015.

27 A. Ess, B. Leibe, and L. Van Gool. Depth and appearance for mobile scene
analysis. In *Proceedings of IEEE International Conference on Computer Vision*,
pages 1–8, Rio de Janeiro, Brazil, October 2007.

28 A. Lerner, Y. Chrysanthou, and D. Lischinski. Crowds by example. *Computer Graphics Forum*, 26(3):655–664, 2007. URL https://onlinelibrary.wiley.com/doi/abs/10.1111/j.1467-8659.2007.01089.x.

29 D. Giustiniano, G. Bianchi, A. Conti, S. Bartoletti, and N. B. Melazzi. 5G and beyond for contact tracing. *IEEE Communications Magazine*, 59(9):36–41, 2021.

30 W. O. Kermack, A. G. McKendrick, and G. T. Walker. A contribution to the mathematical theory of epidemics. *Proceedings of the Royal Society of London. Series A, Containing Papers of a Mathematical and Physical Character*, 115(772):700–721, 1927.

31 E. Pepe, P. Bajardi, L. Gauvin, F. Privitera, B. Lake, C. Cattuto, and M. Tizzoni. COVID-19 outbreak response, a dataset to assess mobility changes in Italy following national lockdown. *Scientific Data*, 7(1):230, 2020.

32 M. H. C. Garcia, A. Molina-Galan, M. Boban, J. Gozalvez, B. Coll-Perales, T. Şahin, and A. Kousaridas. A tutorial on 5G NR V2X communications. *IEEE Communication Surveys & Tutorials*, 23(3):1972–2026, 2021.

33 C. Campolo, A. Molinaro, F. Romeo, A. Bazzi, and A. O. Berthet. 5G NR V2X: On the impact of a flexible numerology on the autonomous sidelink mode. In *Proceedings of the IEEE 5G World Forum (5GWF)*, pages 102–107, Dresden, Germany, October 2019.

34 R. Di Taranto, S. Muppirisetty, R. Raulefs, D. T. Slock, T. Svensson, and H. Wymeersch. Location-aware communications for 5G networks: How location information can improve scalability, latency, and robustness of 5G. *IEEE Signal Processing Magazine*, 31(6):102–112, 2014.

35 A. Conti, S. Mazuelas, S. Bartoletti, W. C. Lindsey, and M. Z. Win. Soft information for localization-of-things. *Proceedings of the IEEE*, 107(11):2240–2264, 2019.

36 M. Z. Win, Y. Shen, and W. Dai. A theoretical foundation of network localization and navigation. *Proceedings of the IEEE*, 106(7):1136–1165, 2018.

37 J. A. del Peral-Rosado, G. Seco-Granados, S. Kim, and J. A. López-Salcedo. Network design for accurate vehicle localization. *IEEE Transactions on Vehicular Technology*, 68(5):4316–4327, 2019.

38 TS 22.261. Service requirements for the 5G system. In *3rd Generation Partnership Project, Technical Specification Group Radio Access Network, Release 17*, 2020.

39 3GPP TR 38.857. Study on NR Positioning Enhancements. Tech. Spec. Group Radio Access Network, Rel-17, 2021.

40 TR 101 607. Intelligent Transport Systems (ITS); Cooperative ITS (C-ITS), February 2020. Release 1.

41 EN 302 665. Intelligent Transport System (ITS) Communications Architecture, September 2010. Release 1.

42 TS 102 637-2. Intelligent Transport Systems (ITS); Vehicular Communications; Basic Set of Applications; Part 2: Specification of Cooperative Awareness Basic Service, March 2011. Release 1.

43 ES 302 637-3. Intelligent Transport Systems (ITS); Vehicular Communications; Basic Set of Applications; Part 3: Specifications of Decentralized Environmental Notification Basic Service, September 2014. Release 1.

44 EN 302 895. Intelligent Transport Systems (ITS); Vehicular Communications; Basic Set of Applications; Local Dynamic Map (LDM), September 2014. Release 1.

45 TR 103 562. Intelligent Transport Systems (ITS); Vehicular Communications; Basic Set of Applications; Analysis of the Collective Perception Service (CPS); Release 2, December 2019. Release 2.

46 TS 103 300. Intelligent Transport Systems (ITS); Vulnerable Road Users (VRU) awareness; Part 3: Specification of VRU awareness basic service; Release 2, November 2020. Release 2.

47 EN 302 890. Intelligent Transport Systems (ITS); Facilities Layer function; Part 2: Position and Time management (PoTi); Release 2, March 2020. Release 2.

48 F. Takaoka. Shibuya crossing as a non-tourist site: Performative participation and re-staging. In *Understanding Tourism Mobilities in Japan*, pages 158–169. Routledge, 2020.

49 5GAA website, 2022. https://www.5gaa.org, Last Accessed: June 17, 2022.

50 5G automotive association Whitepaper, C-V2X use cases volume II: Examples and service level requirements, October 2020.

51 5GAA Report. System Architecture and Solution Development; High-Accuracy Positioning for C-V2X, February 2021.

52 TR 22.872. 3rd Generation Partnership Project (3GPP), Technical Specification Group Services and System Aspects; Study on positioning use cases; Stage 1, September. 2018. Release 16.

7

Location-Aware Network Management

Sergio Fortes[1], Eduardo Baena[1], Raquel Barco[1], Isabel de la Bandera[1], Zwi Altman[2], Luca Chiaraviglio[3], Wassim B. Chikha[2], Sana B. Jemaa[2], Yannis Filippas[4], Imed Hadj-Kacem[3], Aristotelis Margaris[4], Marie Masson[2] and Kostas Tsagkaris[4]

[1] Telecommunication Research Institute (TELMA), Universidad de Málaga, Málaga, Spain
[2] Orange Labs, Châtillon, France
[3] Department of Electronic Engineering, University of Rome Tor Vergata and CNIT, Rome, Italy
[4] Incelligent P.C., Athens, Greece

This chapter investigates the location-based analytics for network management. The relevance and novelties that location-awareness bring to the planning, optimization, and failure management of the cellular networks are detailed, presenting both the general concepts and applicability as well as specific use cases based on it.

7.1 Introduction

New 5th generation (5G) features, such as massive multiple-input-multiple-output (MIMO) and beamforming, multi-connectivity, use of millimeter wave (mmWave) and unlicensed bands, hugely amplify the network operators' need to efficiently manage networks as their complexity increases. To achieve the optimum end-to-end service performance for the growing, highly diverse, and extremely demanding 5G traffic, user-centric network management should replace the traditional network-oriented operation. Here the use of location information of the network users and advanced data analytic techniques is deemed necessary to achieve an enhanced smart network management. Table 7.1 lists the acronyms used in this Chapter.

Positioning and Location-based Analytics in 5G and Beyond, First Edition.
Edited by Stefania Bartoletti and Nicola Blefari Melazzi.
© 2024 The Institute of Electrical and Electronics Engineers, Inc. Published 2024 by John Wiley & Sons, Inc.

Table 7.1 List of acronyms.

Acronym	Definition
5G	5th generation
6G	6th generation
AI	Artificial intelligence
AoI	Area of interest
B5G	Beyond 5G
BS	Base station
CAPEX	Capital expenditures
CCO	Capacity and coverage optimization
CNN	Convolutional neural network
CPU	Central processing unit
CX	Customer experience
DBSCAN	Density-based spatial clustering of applications with noise
DL	Deep learning
DT	Drive tests
EMF	Electro-magnetic field
eNB	eNodeB
FLC	Fuzzy logic controller
gNB	gNodeB
GNSS	Global navigation satellite system
GoB	Grid of beams
HDBSCAN	Hierarchical DBSCAN
KPI	Key performance indicator
LM	Localized measurement
LTE	Long-term evolution
MDT	Minimization of drive tests
MIMO	Multiple-input-multiple-output
ML	Machine learning
MLE	Maximum likelihood estimation
M-MIMO	Massive MIMO
mmWave	Millimeter wave
MR-DC	Multi-radio – dual connectivity
MU	Multi-user
MUT	Mean user throughput

Table 7.1 (Continued)

Acronym	Definition
NCA	Neighborhood component analysis
NSI	Network synthetic image
OPEX	Operational expenditures
OSS	Operations support system
PCA	Principal component analysis
POI	Point of interest
PPCA	Probabilistic PCA
PTS	Power traffic sharing
QoS	Quality of service
RAN	Radio access network
RAT	Radio access technology
REM	Radio environment map
RF	Radio frequency
RGB	Red green blue
RSRP	Reference signal received power
SAPTS	Social-aware PTS
SINR	Signal-to-interference-plus-noise ratio
SLA	Service-level agreement
SNR	Signal-to-noise ratio
SRI	Signal-to-interference ratio
SVM	Support vector machine
TTI	Transmission time interval
UE	User equipment
YUV	(Y) Luma, or brightness, (U) blue projection, and (V) red projection

As it is summarized in Figure 7.1 and it is going to be detailed in this chapter, accurate network localization and advanced network analytics will improve network management throughout the complete network management cycle.

Considering its pre-operation stage, location-awareness will be used for *(re-) planning* purposes, meaning the deployment of network elements, e.g. cell site selection. In the operation stage, the positioning information will be exploited for

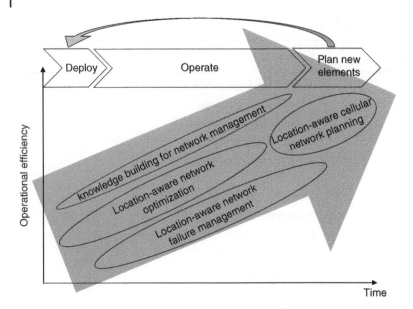

Figure 7.1 Cellular network management functions and location-awareness.

network *optimization* to improve the network performance under dynamic conditions. Moreover, network problems/service degradations require location-aware *failure management* actions to properly detect them, identify their causes and solve them.

Classical network management approaches typically rely on two types of data sources:

- Cell-level metrics: Performance/radio measurements' statistics per each observation period (e.g. hourly dropped call rate, daily average reference signal received power (RSRP)). This type of metrics (together with alarms/events) are the classic inputs for inference mechanisms applied to detection, diagnosis, and control. These metrics lack localization data and, therefore, they do not take advantage of the improvement that location can bring in the management of cellular networks [1–3].
- Drive tests and minimization of drive tests (MDT) traces: user equipment (UE) level traces, localized via global navigation satellite system (GNSS) or cellular localization techniques. Although they provide a huge amount of information, they are difficult to adapt to existing automated methods and are commonly analyzed directly by human experts.

In this way, individual UE traces might be difficult to be applied to inference mechanisms, and they might require preliminary stages of feature engineering [4] and clustering. In this way, *knowledge building* through processing of the data shall support all the management activities, being executed in parallel in order to provide location-enriched analytics. Thus, new knowledge is built by integrating different variables (e.g. location and performance), the construction of novel metrics, the identification of point of interests (POIs), or the forecasting of the traffic demand or other key performance indicators (KPIs).

In this chapter, the relevance and novelties that location-awareness bring to the planning, optimization, and failure management of the cellular networks are detailed, presenting both the general concepts and applicability, as well as specific use cases based on it.

7.2 Location-Aware Cellular Network Planning

In this section, we shed light on the impact of localization on the planning of cellular networks. First, we provide a general overview about the cellular network planning problem which can be useful for the non-experts in the field. Second, we motivate the introduction of localization in the planning phase. Third, we provide the high-level formulation of the location-aware cellular network planning problem. Fourth, we discuss the future directions in the field.

7.2.1 What Is the Cellular Network Planning?

The cellular network planning is a key task of mobile network operators [5]. The planning of a given mobile technology requires, in fact, the selection of the sites to host the new radiating antennas, as well as the configuration of the parameters of each installed panel. The main goal of the planning phase is to provide an adequate signal coverage over the considered territory, while at the same time ensuring good quality of service (QoS) level for the users. However, the cellular planning is largely governed by rather orthogonal objectives and constraints. First of all, the operators aim at limiting both capital expenditures (CAPEX) and operational expenditures (OPEX) expenses. Therefore, the selection of the sites to install base station (BS) equipment is heavily constrained by the need of reducing as much as possible the associated installation and management costs. In addition, another big constraint of the planning phase is represented by the electro-magnetic field (EMF) levels that are radiated by the installed equipment. In more detail, the operator has to ensure that the composite exposure that is radiated by both antennas to be installed and by the already-deployed equipment (e.g. legacy generations and/or other operators providing mobile service over the same area) fulfills the maximum exposure limits that are imposed by law.

Finding a good balance between the aforementioned objectives and constraints is very challenging, and the operator typically employs a *divide-et-impera* approach, in which the cellular planning is divided in the following steps:

(P1) Selection of the candidate sites that can host the new panels. These sites can be either already-available installations or new sites that have to be created *ad-hoc*;

(P2) Configuration of each site in terms of equipment to be installed, number of sectors, sector horizontal orientation, mechanical and electrical vertical tilting of the panel, maximum radiated power of the panel;

(P3) Requests of the authorizations to install the new panels on the selected sites. The authorizations typically include reports done by third-party companies that perform the simulation of the EMF levels over the territory, including both the already-installed equipment (or by measuring the baseline exposure over the critical locations) and the numerical levels generated by the new antennas in (P1) and configured as in (P2);

(P4) Installation and activation of the new antennas in the sites after being authorized;

(P5) Measurement of the signal level that is radiated by the new antennas over the territory;

(P6) Further optimization of the antennas parameters (such as variation of output power and/or vertical tilting), in order to improve coverage and/or reduce EMF levels (by eventually updating the authorization requests of P3).

7.2.2 Why Is Localization Important in the Planning Phase?

5G antennas heavily employ beamforming and MIMO functionalities [5]. Although both features were already implemented in previous generation antennas (e.g. long-term evolution (LTE)-Advanced), their full exploitation will be realized in 5G and beyond-5G equipment. In more detail, the main idea of beamforming is to concentrate the output power on the zones where the users are located, by synthesizing narrow beams that are pointed toward selected locations. In this way, the spatial reuse of the radio resources is highly increased compared to the non-beamforming case, and the interference on the signal channel is also reduced. On the other hand, the idea behind MIMO is to exploit multiple antenna elements (both in transmission and in reception) to take advantage of the multi-path propagation and increase performance to users.

Apart from improving QoS levels, both beamforming and MIMO have a huge impact on the EMF levels over the territory [6]. In fact, the large exploitation of such techniques can result in a huge decrease of EMF levels radiated by the BS (compared to previous generations), mainly because the output power is not uniformly distributed over the territory, but it is highly optimized to cover only the users that require the mobile service.

In the extreme case, the beams from the base station may be synthesized only toward the active users, and therefore the knowledge of the user positioning becomes a fundamental information from the operator side. Although current base stations mainly implement "static" features, in which the beams are statically synthesized toward pre-defined positions, it is expected that future deployments, exploiting higher-than-mmWave frequencies, will have to aggressively apply "dynamic" beamforming strategies, able to serve the users over the given area with dedicated beams.

In this scenario, the localization accuracy will play a great role in determining both the level of throughput and EMF, thus in turn heavily influencing the planning phase [6, 7]. Intuitively, when the localization accuracy is coarse, the beams tend to be more overlapped over the territory. This condition tends to increase the EMF levels and to decrease the throughput (due to increase of interference). On the contrary, when the localization accuracy is very fine, "pencil" beams are synthesized, with a notable decrease of EMF levels and increase of throughput.

To provide evidence of the previous observations, we numerically compute the EMF and throughput by simulating a dynamic antenna panel system, able to tune the beam orientation and beam widths in accordance to the precision in localizing the served users. We refer the reader to [6] for the technical details, while here we provide the salient outcomes. In brief, Fig 7.2 reports the electric field in V/m over the coverage area of an omnidirectional dynamic antenna panel. The sub-figure on the left is the field that is experienced when a localization accuracy of 20 m is assumed, while the sub-figure on the right reports the same scenario with a lower

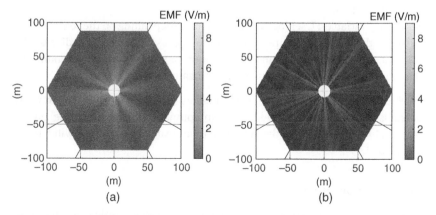

Figure 7.2 EMF levels (in V/m) for two different values of localization accuracy and the same subset of user positioning. The EMF for the localization accuracy equal to 2 m (b) is clearly lower than the 20 m case (a) [6].

localization accuracy (2 m). Interestingly, the EMF levels are greatly reduced when passing from 20 m to 2 m of localization accuracy, as the overlapping of the beams is greatly reduced. This effect is also beneficial for the throughput levels, which is improved when the localization accuracy is reduced (details in [6]).

Obviously, EMF and throughput have a great influence of the planning phase, both in terms of coverage levels and compliance against the maximum limits defined by law [8]. Eventually, the localization quality is also affected by the positioning of the base stations themselves. In particular, the more antennas are installed on a given territory, the better is the localization accuracy, thus again positively influencing both EMF and throughput levels.

7.2.3 Location-Aware Cellular Network Planning

Having understood the importance of localization in the process of cellular network planning, we shed light on an innovative planning formulation, taking into account the location accuracy of the users. More formally:

- Given: Positioning of the candidate sites, antenna configurations for the equipment already installed over the territory, positioning of the users, antenna features of the panels to be installed;
- Minimize: CAPEX and OPEX costs of the installed equipment;
- Subject to: Coverage constraints, throughput guarantees, EMF compliance against the limits defined by law, operator own-policies, physical impairments of the antenna panels (such as maximum number of beams);
- Control variables: Site selection, number of sectors, panel orientation (horizontal/vertical mechanical/electrical tilting), maximum radiated power for each antenna;
- Secondary variables: User localization accuracy (depending on the selected sites as well as the panels configuration), beam-to-user association (depending on the selected sites as well as the panels configurations), beam widths (depending on the beam-to-user association and the user localization accuracy).

Obviously, it is expected that the mathematical formulation of the previous problem is highly non-linear, and hence sub-optimal solutions, based, e.g. on the constraint linearization or on the definition of innovative algorithms, will have to be developed.

7.2.4 Future Directions

The integration of localization in the cellular planning phase could boost the adoption of highly dynamic antenna panels, able to synthesize traffic beams and MIMO flows only toward the served users. This scenario is particularly promising when considering mmWave and higher frequencies, which are subject to strong propagation impairments compared to sub-6 GHz frequencies. In addition, the current densification trend for the deployment of mobile networks [8] is a natural driver for improving the localization accuracy and hence both throughput and EMF levels. A cellular planning assisted by localization, in fact, could lead to a reduction of CAPEX and OPEX costs for the operator, while at the same time improving the throughput to users and decreasing the overall EMF levels over the territory.

A second important innovation brought by a location-aware planning is in the compliance of assessment procedures, which are used to verify whether the numerical EMF adheres with the limits defined by the law. Although state-of-the-art compliance procedures (see, e.g. [9]) are still based on over-simplifying assumptions, like free space propagation loss, antenna diagrams based on the envelope of beams, and a radiated power from the panel always equal to the maximum value, there is ample room in revisiting those guidelines when dynamic antenna panels, working on mmWave or higher frequencies, are employed. In this domain, the application of more realistic propagation models, taking into account the buildings and their effects on propagation, is a promising area of research, and one of the proposals currently under evaluation in the relevant committees working on the compliance of assessment standards. In addition, the application of location-aware EMF computations could improve also the quality of the numerical evaluation of the predicted field levels, which could better match the actual ones that are then measured during the operation phase. This step is particularly important in countries (like Italy) that are enforcing more stringent EMF limits than the ones defined in international regulations [8], thus (possibly) allowing the installation of antenna panels that could be otherwise denied when legacy compliance of assessment procedures is applied.

7.3 Location-Aware Network Optimization

7.3.1 What Is the Cellular Network Optimization?

The term cellular network optimization encompasses all tools, techniques, and/or even best practices that are used to improve network performance. It implies an ongoing process of optimization and not a one-time action taking place. At the same time, network optimization refers to one or more KPI with respect to which the network is to be improved. The latter inherently implies that the said KPIs are measured and monitored in a continuous manner, and a closed loop process (for automated network optimization) exists where changes in a network lead to new KPI values and KPI values are improved through suggested optimization actions.

These KPIs may refer to those measured at a UE level or at network level enhancing QoS and customer experience (CX), as well as business objectives and service-level agreements (SLAs).

Based on the above, a cellular network optimization problem should take into account the following:

- The optimization objective, i.e. one or more specific network measurements that are monitored and are targeted to be improved;
- The granularity of the optimization problem, and specifically where and when to implement this optimization process;
- The relevant action space, meaning the set of actions and/or network parameters that can be changed through an action in order to optimize the network;
- Any and all regulatory, business and/or system/equipment (incl. hardware and software) limitations that would reduce the action space or limit the actions in terms of frequency and range of implementation.

7.3.2 Why Is Location Information Important in Optimization?

In this context, the inclusion of UE positioning data and related advanced analytics and machine learning (ML) or artificial intelligence (AI)-based results can be of key importance as it adds an extra layer of information that can drive to further insights. Indicatively, specific areas, such as entertainment venues/concert halls, stadiums (and surrounding areas), shopping malls, and university campuses, may display highly irregular, and/or complex population densities, traffic demand patterns, or even velocity patterns compared to average rural areas.

These factors may render other techniques to underperform. However, by utilizing more advanced analytical solutions that can leverage mobile terminal location information, we can pinpoint more clearly for example specific areas of network quality degradation at a spatio-temporal level, e.g. to a specific sub-area/POI or to a specific path/route on the map that displays higher traffic demand or diverse velocity profiles at different times in a day, week, season, and so on.

Figure 7.3 exemplifies how cellular network optimization utilizing location information can take place in an automated manner assuming a case where specific parameter settings can be configured in an automated way, such as configuring the optimal azimuth angle setting (see Sections 7.3.3 and 7.3.4). Briefly, for a set of cells in a predefined area, the radio access network (RAN) Operations Support System (OSS) monitors network parameters like RSRP, signal-to-interference and noise ratio (SINR), etc. At the same time, location information (related also with site positions) allows for the detection of problematic sub-areas and corresponding cells either real-time or at regular time intervals. An optimization module proposes the optimal setting based on the aforementioned information

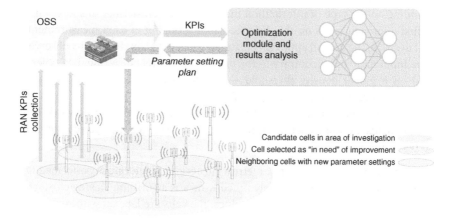

Figure 7.3 Network optimization and location information.

utilizing either a specific algorithm or simulations and the ML-based predicted impact these settings will have in order to improve network performance.

7.3.3 Hybrid Clustering-Based Optimization of 5G Mobile Networks

As mentioned previously, location information from moving UEs within a predefined area can be used for exploratory purposes, such as detecting POIs, i.e. zones of high UE density, and high/low signal to interference and noise ratio (SINR), using ML. By building knowledge for the optimization of 5G capacity layers in modern urban environments and achieving the optimum configuration of the 5G gNodeB (gNB) cell parameters, i.e. the maximization of achieved coverage RSRP and quality SINR of the served mobile terminals for very high data transmission. Such an example is provided by [10], where authors apply various distance- and density-based clustering techniques [11] on spatial (UE position) information for POI detection, and combined with network KPI measurements, performance of this hybrid approach is tested. More specifically, the enhanced hybrid clustering results are tested using an algorithm for azimuth steering against the clustering results from (i) spatial-based clustering and (ii) network-based clustering assuming an original network setup in a predefined simulation environment, aka "playground."

7.3.3.1 Clustering Methods and Algorithmic Approach
The main clustering methods employed involved the distance-based K-means and the extension of the density-based spatial clustering of applications with noise (DBSCAN) algorithm called hierarchical DBSCAN (HDBSCAN) [11, 12],

depending on the type of information included in the clustering. In short, while the first approach relies on the distance-based grouping of each measurement/point in space, the second approach creates clusters based on the density observed in the input data and – contrary to the original DBSCAN clustering data objects based on a global density threshold – it generates a complete density-based clustering hierarchy composed only of the most significant clusters. In order to prove the enhanced approach of using both UE positions and network KPI measurements, authors test three different approaches, incl. the hybrid approach and two baseline scenarios:

- Baseline approach 1 – Network metrics clustering: Clustering is performed on network KPIs using appropriately tuned K-means algorithm. Having clustered the UEs based on RSRP and SINR values, the cells are labeled with respect to their KPI scores of the clusters that are serving. More specifically, a cell with low score for both is labeled as "bad" and the derived action is to rotate its closest "good" cell to the direction of the geometric center of the area covered by the original "bad" cell to improve upon their RSRP and SINR scores.
- Baseline approach 2 – Spatial clustering: UEs are clustered using only their coordinates with K-means and – as in baseline approach 1 – each gNB is associated to a subset of the clusters and is rotated toward the geometric center of these clusters.
- Hybrid clustering approach: UEs are clustered on both coordinates and network measurements with HDBSCAN chosen based on the nature of the data objects. Then, similar to the network approach, each gNB is characterized by the network scores of its associated cluster. Finally, when a "bad" cell is found, it is rotated to center its direction with the geometric center of its associated clusters such as in the spatial method.

7.3.3.2 Results and Conclusions

Figure 7.4 presents the cluster centroids derived from the hybrid clustering scheme in (a), and the results in the simulation playground from azimuth steering actions following the aforementioned clustering (b).

To verify the applicability of this hybrid approach, the clustering results were compared to the baseline approaches described in Section 7.3.3.1 and with respect to the results of azimuth steering. Since the maximization of coverage and quality is targeted, authors compared the RSRP and SINR metrics of the total cellular network area under investigation. More specifically, both the cumulative distribution functions (results not shown) as well as the 25th, 50th, and 75th percentiles were employed for quantitative results.

Figure 7.5 presents such a comparison of the distribution of percentiles of the data generated by all the different reconfiguration schemes for all evaluation KPIs,

        (a)   (b)

Figure 7.4 (a) Visualized cluster centroids and (b) simulation playground following hybrid clustering and azimuth steering actions [10] Margaris et al., 2022/MDPI/CC BY 4.0.

i.e. RSRP gains and SINR gains compared to the initial azimuth angles setup where no action takes place.

Overall, as shown in the figure, all algorithmic approaches displayed an improvement compared to the original network setup, which indicates the value of exploiting ML- and/or AI-derived results for performing "informed" network optimization actions. It also implies that location data can improve upon decisions for network optimizations as originally anticipated. In addition to the above, both baseline approaches had a similar performance. However, the proposed scheme outperformed the baseline algorithms for the investigated simulation playground by a factor of more than 100% in most cases, as shown in Figure 7.5, providing concrete evidence for the approach's applicability and the use of positioning-related data in network optimization schemes.

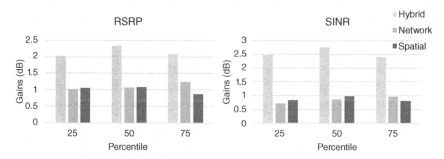

Figure 7.5 RSRP (a) and SINR (b) gains comparison for hybrid clustering vs. baseline approaches [10].

7.3.4 Location-Aware Capacity and Coverage Optimization

Location information can be used to improve capacity and coverage optimization methods. In this case, the proposed approach is applied to a realistic dense urban scenario in which the eNodeBs (eNBs) and gNBs are co-located in the same sites but having different configuration. In this regard, a location-aware capacity and coverage optimization (CCO) system is proposed to maximize the capacity of users located in ultra-dense areas as well as to guarantee an optimal quality of service of other users. Moreover, this capacity optimization is carried out while a previous optimization of the coverage throughout the whole scenario is maintained. The proposed methodology optimizes the configuration of the eNBs and gNBs in multi-radio – dual connectivity (MR-DC) scenarios.

7.3.4.1 Dual-Connectivity Optimization

The method consists of two main parts, which are performed one after the other: coverage optimization phase and capacity optimization phase, respectively. The first part focuses on maximizing coverage and minimizing interference level throughout the scenario. The latter part aims at boosting the available capacity for UEs in crowded regions, while the negative impact on the rest of UEs is minimized. At the beginning of the first phase, the scenario is divided into 20 m side size squares. Then, the SINR is computed at the center of each square for each radio access technology (RAT). This allows to identify the areas receiving weak signal or with high interference level. Afterward, the network operator defines a minimum required SINR threshold to detect the squares of each RAT with poor signal. However, only the squares with poor signal in both RATs are considered in the first phase of the framework. Afterward, the centroids of the areas, which consist of bordering squares with poor signal, are computed. The two interfering eNBs and the two interfering gNBs that provide with highest RSRP to each centroid are then identified. Finally, the azimuth and tilt of the main eNB and gNB that will serve the area with low SINR are modified to point at it. In addition, the transmit power is modified based on the distance between the centroid and the node. On the other hand, the parameters of the interfering node are modified to decrease its signal on the mentioned square.

This configuration maximizes global coverage, but the location of the UEs is not used at this point. In this sense, the second phase modifies this configuration to increase the bandwidth that can be allocated to UEs located in crowded regions, such as city centres, taking advantage of multi radio - dual connectivity (MR-DC). Therefore, the location of UEs is utilized to identify the topology of these ultra-dense areas. Then, the second algorithm identifies the corresponding eNBs and gNBs that provide them with the strongest signal. In addition to changes in the previous antenna parameters of these nodes, this second stage modifies the threshold of the event A2 to ease users leaving crowded areas to be served by external nodes. This enables increasing the resource availability for UEs in these areas. It should be pointed out that this threshold modification only applies to gNBs, since eNBs are used as connection anchor to decrease the probability of dropping.

7.3.4.2 Results and Conclusions

Figure 7.6 represents the variance of the throughput gain obtained by the first stage. It shows that the highest guaranteed gain is achieved for the lowest percentiles, i.e. UEs located in areas of low coverage or high interference. The average of this gain is about 15%. Then, Figure 7.7 shows the throughput gain reached by the second stage with respect to the performance obtained by the previous phase

Figure 7.6 Throughput gain after the first phase for users throughout the scenario.

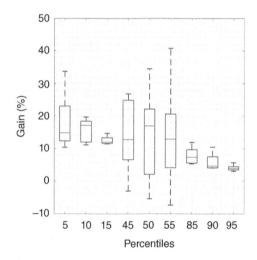

Figure 7.7 Throughput gain after the second stage for users in normal areas.

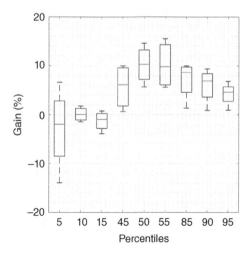

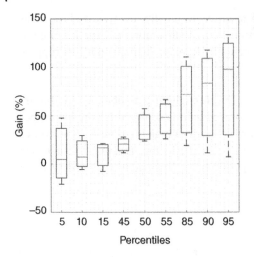

Figure 7.8 Throughput gain after the full framework for users in crowded areas.

in areas which have not been considered as ultra-dense. The results show that the achieved throughput hardly varies for the low percentiles, as the usual values are close to 0% gain. However, it should be noted that the gain of the 5th percentile may decrease to 10% in some scenarios where the parameters of the same nodes reconfigured by the first phase are changed to cover a crowded region. Despite this possible decrease, the final gain is always positive with respect to the initial performance before using the proposed framework.

Finally, only users located in crowded areas are analyzed to accurately evaluate the enhancement achieved by the second stage in Figure 7.8. It depicts the throughput gain obtained by the complete framework only for UEs located in ultra-dense areas compared to the performance before using any stage. It shows that the throughput gain gradually grows as the users have better radio conditions and higher resource availability for them. On the other hand, there is a low probability that users in worst conditions deteriorate their throughput, as indicated by the 5th percentile. This corresponds to users leaving crowded areas, which are poorly served by external nodes when forced by the event A2.

7.3.5 SINR Prediction in Presence of Correlated Shadowing in Cellular Networks

SINR (Signal-to-Interference-plus-Noise Ratio) is radio indicator used in mobile network planning tools to calculate the coverage rate and to estimate the data throughput based on the Shannon formula. Hence, a good estimation of this metric at any user location ensures proper planning and optimization of radio coverage and good estimation of end-user throughput. Predicting the SINR without taking into account the error in its estimate does not allow the operator to know the accurate rate perceived by the user. The main objective of this study is to propose a method of predicting SINR values, as well as the variance of the error of its estimate. Once this estimation is set up, the operator can predict the average throughput per unit of resource served to a user in a given location. This work proposes the combination of the Kriging technique for received signal prediction based on geolocalized measurements described in [13] and a previous work described in [14] in order to establish the distribution of the predicted SINR. Initially, Kriging was used in mobile networks to predict a received power value by spatially interpolating geo-located measurements [15].

The present functionality goes beyond previous works by considering the prediction error of received signals from serving and interfering cells. We give the predicted SINR and its variability in a new location for the cases where the cells are assumed to be independent and where they are assumed to be correlated. When cells are assumed to be correlated, we take into account both the spatial correlation of shadowing signals for each cell and also the inter-cell correlation by considering the model for generating multiple links shadowing samples proposed in [16].

7.3.5.1 SINR Prediction with Kriging

Let a given UE located at location $x \in \mathbb{R}^2$. Taking into account that cells can use the same frequency band as it is the case for the long term evolution (LTE) network, the UE receives a useful signal from a serving cell k and M interfering signals from M neighboring cells numbered from 1 to M (note that $k \notin \{1, \ldots, M\}$). Its measured received signal power from cell i (either serving or interfering cell) in the logarithmic domain is given by [15]

$$z_i(x) = P_i^{(t)} - 10\alpha \log(d_i(x)) + s_i(x) + \varepsilon_i(x), \tag{7.1}$$

where $P_i^{(t)}$ is the transmit power of the ith cell which includes the cell power, antenna gains, and feeders losses, $d_i(x)$ is the normalized distance between the considered UE and the ith cell, α is the path loss-exponent which depends on the propagation environment [17]. $s_i(x)$ follows a zero-mean Gaussian random variable with variance ω_i^2 that indicates the log-normal shadowing [18] and $\varepsilon_i(x)$ is a centered uncorrelated Gaussian random variable with variance σ_i^2 that models any error in measurement. We assume that $s_i(x)$ and $\varepsilon_i(x)$ are independent random variables.

The aim of this study is to predict the SINR for a given location x_0 based on some measurements $z_i = (z_i(x_1), \ldots, z_i(x_N))^T$ with $(\cdot)^T$ denoting the transposition, reported by N user equipments located at N different locations $x_1, \ldots, x_N$. The received signal powers $z_i(x_m)$ and $z_i(x_n)$ measured at locations x_m and x_n are correlated because the shadowing signals $s_i(x_m)$ and $s_i(x_n)$ are correlated. The correlation is increasingly strong when the locations x_m and x_n are close. Several studies express the correlation between received signals as a function of the distance separating the measurement positions [19, 20]. In this work, we consider the *spherical* model of the covariance matrix [20] according to which the covariance function is

$$c^{(sph)}(x_m, x_n, \omega_i, \tau) = \begin{cases} \omega_i^2 - \omega_i^2 \left(1.5 \frac{|x_m - x_n|}{\tau} - 0.5 \frac{|x_m - x_n|^3}{\tau^3}\right) & \text{if } |x_m - x_n| \leq \tau \\ 0 & \text{otherwise.} \end{cases}$$

(7.2)

where and τ is the typical autocorrelation decay distance. The corresponding covariance matrix is $\Sigma_i = C_i + \sigma_i^2 I_n$, where $\Sigma_i(m, n) = \text{cov}(z_i(x_m), z_i(x_n))$, $C_i(m, n) = c^{(sph)}(x_m, x_n, \omega_i, \tau)$. We notice that Σ_i depends on ω_i, τ, and σ_i. These parameters are unknown in reality and need to be estimated. Their respective estimators $\hat{\omega}_i$, $\hat{\tau}$, and $\hat{\sigma}_i$ are obtained by maximum likelihood or by the method of moments [15]. In simulations, we will use the maximum likelihood estimation (MLE) method [16] for estimating Σ_i.

In order to predict the SINR at a new location x_0, we first predict the received signal power at x_0 using the Kriging interpolation technique [13]. By denoting $v_i(x_0)$ the received signal power at the location x_0 without the measurement error, the corresponding Kriging predictor is [13]

$$\hat{v}_i(x_0) = y_i^T(x_0)\hat{\beta}_i + \hat{c}_i^T(x_0)\hat{\Sigma}_i^{-1}(z_i - Y_i\hat{\beta}_i),$$

(7.3)

where $\hat{c}_i(x_0) = \left[c^{(sph)}(x_0, x_1, \hat{\omega}_i, \hat{\tau}), \dots, c^{(sph)}(x_0, x_N, \hat{\omega}_i, \hat{\tau})\right]^T$ denotes the estimate of the shadowing covariance vector between the considered location x_0 and the observations, $y_i(x_0) = (1, -10\log(d_i(x_0)))^T$ and $\hat{\Sigma}_i$ is the estimator of Σ_i. $\hat{\beta}_i$ is the generalized least square estimator of $\beta_i = (P_i^{(t)}, \alpha)^T$ [21].

Now, we propose to inject the predicted received signals powers into the expression of SINR as following to obtain the prediction of the SINR at the new location x_0 [16]

$$\widehat{SINR}_{Kr}(x_0) = \frac{10^{\frac{\hat{v}_k(x_0)}{10}}}{\zeta + \sum_{i=1}^{M} \rho_i 10^{\frac{\hat{v}_i(x_0)}{10}}}.$$

(7.4)

$\widehat{SINR}_{Kr}(x_0)$ does not take into account the errors of the estimates of the received signal power. To evaluate the impact of prediction errors on the SINR, we introduce the following random variable [16]

$$\widehat{SINR}_{RV}(x_0) = \frac{10^{\frac{\hat{v}_k(x_0) + \xi_k(x_0)}{10}}}{\zeta + \sum_{i=1}^{M} \rho_i 10^{\frac{\hat{v}_i(x_0) + \xi_i(x_0)}{10}}},$$

(7.5)

where $\xi_i(x_0)$ is a zero-mean Gaussian random variable with variance $\gamma_i^2(x_0) = E[\{v_i(x_0) - \hat{v}_i(x_0)\}^2]$. $\gamma_i^2(x_0)$ is the mean square prediction error of the received signal power. In order to take into account the inter-cell correlation, we suppose that the correlation coefficient between $\xi_i(x_0)$ and $\xi_j(x_0)$ for $i \neq j$ (i and j can be either serving or interfering cell) is equal to the shadowing correlation coefficient of the two paths i and j. Note that $\widehat{\text{SINR}}_{\text{RV}}(x_0)$ measures the spread of the values of the SINR at x_0, given the observed values of z_i at locations $x_1, \ldots, x_N$. This allows to calculate the probability of coverage and predict the average rate of a UE at a new location x_0.

According to [14], $\widehat{\text{SINR}}_{\text{RV}}(x_0)$ is a lognormal random variable. We determine its characteristic parameters in the logarithmic domain following the same method proposed in [14, 22].

7.3.5.2 Results and Conclusions

For the validation, we consider a Monte Carlo simulation of a mobile network where cell positions (serving and neighbors) are obtained from a real implementation of an LTE network located in East France. We consider an area of 155m × 155m surface to perform the study, where user locations are regularly spaced on a Cartesian grid consisting of 5 × 5 m squares. This yields a total of 1024 measurement points.

Figure 7.9 gives the predicted SINR on test locations for $M = 4$ interfering cells. All shadowing signals of serving and interfering cells are assumed to be correlated

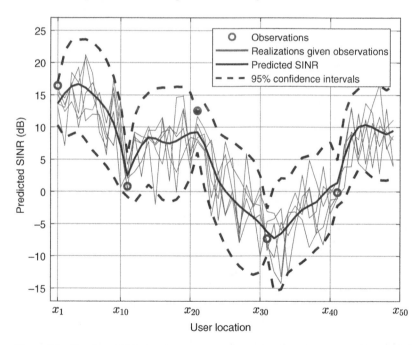

Figure 7.9 Predicted SINR for $M = 4$ interfering cells, a standard deviation of shadowing signals of 10 dB, a noise variance of 3 dB, a correlation coefficient of 0.5, and a training ratio of 10%.

with a coefficient of correlation of 0.5. All cells have the same standard deviation value of 10 dB. The variance of noise is set to 3 dB. The ratio of the training locations is 10% of the whole dataset. The circles in Figure 7.9 correspond to observations. The thick solid curve gives the Kriging predictor given the observations, while the thick dotted ones correspond to 95% confidence intervals which are calculated from the error standard deviation of the prediction. The thin curves give some possible realizations of the true SINR. That means that all thin curves are predicted with the same function using Kriging method (thick curve). From Figure 7.9, we show that it is important to consider both the predicted SINR and its prediction error. In fact, we show that the predictor may overestimate the SINR, and then the mobile network may allocate fewer resources than the user would need to satisfy the required data rate. Note that the prediction standard deviation is non-null at the observation points due to measurements errors. In [16], SINR prediction error has been shown to not impact the predicted average data rate perceived by the end user when the number of training positions is high enough.

Figure 7.10 shows the probability of coverage for two test locations. When we do not take into account the error in prediction, the predicted SINR with the Kriging technique is 13.13 dB for the first test location and 10.78 dB for the second test

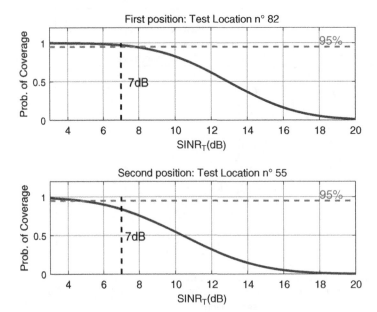

Figure 7.10 Coverage probability for $M = 4$ interfering cells, a standard deviation of shadowing signals of 10 dB, a noise variance of 3 dB, a correlation coefficient of 0.5, and a training ratio of 10%.

location. We assume that the operator requires that the probability of coverage must be greater than 95% with an SINR threshold of 7 dB to ensure a satisfactory QoS for its users. In this case and taking into account the variability of the SINR because of the prediction errors, we can deduce that the first position is considered as covered by the service and the second location is considered as not. Thus, the operator must invest to correct the coverage problem and satisfy its customers in location 2. However, if we have only used the Kriging predictor without taking into account the error in prediction, the two locations would be supposed as covered since 13.13 dB and 10.78 are both greater than SINR threshold (7 dB).

7.3.5.3 Multi-user (MU) Scheduling Enhancement with Geolocation Information and Radio Environment Maps (REMs)

Massive multiple-input-multiple-output (M-MIMO) and beamforming achieve higher gains for users by focusing the energy on them. M-MIMO also reduces intra-cell interference. Users located at the edge of the cell may however experience high interference from the neighboring cells. To manage interference between neighboring cells, we propose a low-complexity solution with a coordinated MU scheduler that leverage radio environment map (REM) knowledge and user location to optimize the resource allocation. The REMs provide the RSRP maps for the Grid of Beams (GoB) of the serving cell and the neighboring interfering cells. The quality of the REMs in terms of resolution and precision is critical for the derivation of correct coordination decisions. To this end, Kriging with the covariance tapering spatial interpolation [20, 21] has been used that provides good prediction accuracy and computational complexity. It is assumed that the measurements reported to the network include the identity of the serving beam and that the antenna gain for each beam is known. Consider a system model with two BSs m and m' having each a set of 16 beams b_i. For each b_i, we construct a distinct REM i.

The multi-user scheduling is an iterative process that aims at avoiding the simultaneous scheduling of users that are highly interfered by other's serving beam. We denote $\mathcal{U}_1$ and $\mathcal{U}_2$ the sets of active users attached to the cells m_1 and m_2, respectively. $\mathcal{U}_{k_1}$ and $\mathcal{U}_{k_2}$ stand for the sets of UEs to be scheduled in the new transmission time interval (TTI) by m_1 and m_2, respectively. $\mathcal{U}_{k_1}$ and $\mathcal{U}_{k_2}$ are first initialized to empty sets and are meant to contain at most k_1 and k_2 users, respectively. The users from the sets $\mathcal{U}_1$ and $\mathcal{U}_2$ are first ranked according to a proportional fair criterion. The highest-ranking user, respectively, u_1 and u_2 from $\mathcal{U}_1$ and $\mathcal{U}_2$ are selected. The REMs provide the scheduler with expected signal and interference for both users. The expected values are used to compute the signal-to-interference ratios (SIRs) involving only the beam of the other user. If the minimum of these two SIRs is below a given threshold Γ (decided once with an exhaustive search), only the user whose cell has the priority at this TTI will be scheduled. For instance, if the TTI is even, only the user attached to m_1 will be served, and if the TTI is odd, then only the user attached to m_2 will be served. Once a user is scheduled, the users attached to the same beam or the adjacent are removed from the set of candidates, to prevent inter-cell interference. Moreover, every user belonging to the neighboring cell fulfilling with a minimum SIR above Γ will also be removed. Note that REMs are assumed to be constant for a long period of times as propagation shows little variation over time. This assumption is accounted for by the fact that REMs are here used for signal and not SINRs. Interference is computed dynamically based on REMs and resource allocation later on. Therefore, the only measurement required from users are their location. Consequently, the algorithm does not trigger latencies in the scheduling process.

7.3.5.4 Results and Conclusions
Experiments lead in the network introduced in Figure 7.11 to assess significant improvement of the MUT, with a gain up to 120%. The results of the experiments are presented in Figure 7.12- where the UEs outside the hotspot area are circled in

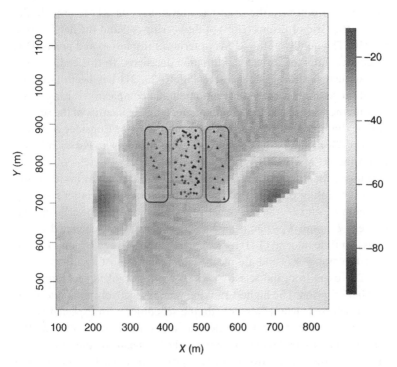

Figure 7.11 Spatial distribution of UEs in cells *m* and *m'*.

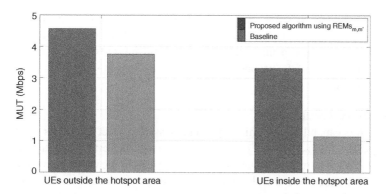

Figure 7.12 mean user throughput (MUT) inside and outside the hotspot areas.

Figure 7.11 by the black rectangles where the UEs located inside the hotspot area are depicted by the points inside the central rectangle. Users inside the hotspot are the most interfered and therefore benefit the most from the collaboration.

7.3.6 Social-Aware Load Balancing System for Crowds in Cellular Networks

Location information is extremely important in the identification of crowds; this means clusters of users that might overload the cellular network. Also to the association of such user concentrations with specific network elements and obtained performances. Novel load balancing systems, as the one to be described, can take advantage of crowd-position knowledge to improve the network service.

7.3.6.1 Social-Aware Fuzzy Logic Controller (FLC) Power Traffic Sharing (PTS) Control

In this area, one approach to the above is based on integrating a fuzzy logic controller algorithm with social-awareness, which considers the relative position between cell sites and the social event venue in order to configure the network parameters [23]. This approach is evaluated for different configurations of load balancing methods simulated on an urban macroscenario, mitigating the impact of the number of users per cell without degrading the signal quality.

As shown in Figure 7.14, the functionality is based on a load balancing system based on fuzzy logic controller (FLC). This is composed of a classic load balancing algorithm (power traffic sharing (PTS)) with the addition of a social-aware load balancing algorithm (social-aware PTS controller (SAPTS)) to achieve a better-balanced load sharing. This is performed by using social-aware input parameters to fine tune the load balancing response of FLC based on inputs from the network and UEs. This mechanism allows to decrease the number of users in the most loaded cell and increase the number of users in adjacent cells, adapting the coverage area of each cell based on social event data. This contribution also maintains adequate values of mean SINR per cell and presents a comparison regarding the adaptation of the proposed mechanism to different venue-locations respect to network cell location.

7.3.6.2 Results and Conclusions
To showcase the capabilities of this approach, a system level simulator has been developed in MATLAB based on an Urban Macro-Cell Test Environment, describing a social event from the concentration of the UEs in the venue before its beginning until the social event ends, and assistants disperse. Users' crowds have been simulated under two different venue locations on a cell network deployment scenario with 5G radio parameters. The sites were located in regular hexagonal pattern following the evaluation configurations for Urban Macro test environments. Some features like inter-site distance were specified with 500 meters between each site with a cell layout of 19 macro site and 3 cells per site, forming a hexagonal grid. Figure 7.14 shows the SINR map from the simulation scenario with 19 sites giving a total of 57 cells where the venue has been located at cell center point (700, 600) mostly covered by cell 5, or cell edge point (700, 600) located in a coverage hole between cells 5, 7 and 1.

A comparison is made between the results obtained in two venue positions (edge and center, as represented in Figure 7.14). Figure 7.15 shows the overload behavior of the cells involved in the social event with the venue located in the cell edge for different configurations:

- No optimization (left figure): without any load balancing the cell 5 is highly overloaded by the event.
- Cellular-data only optimization (center): here the PTS controller of Figure 7.13 runs without the SAPTS part. Two options for the changes of power guided by

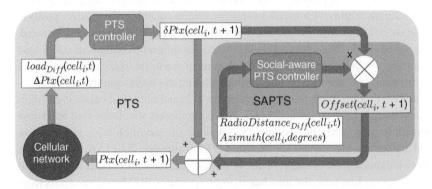

Figure 7.13 Load balancing system based on SAPTS and PTS [23].

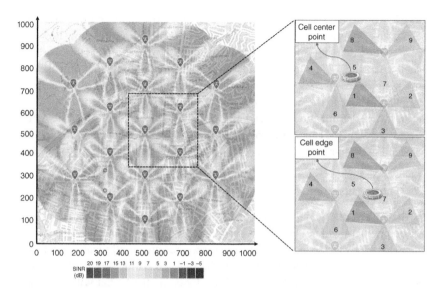

Figure 7.14 SINR map from regular hexagonal 5G Urban Macro scenario using "close-in" propagation model and half-wavelength rectangular microstrip patch antenna element in an 8-by-8 array of 57 cells [23].

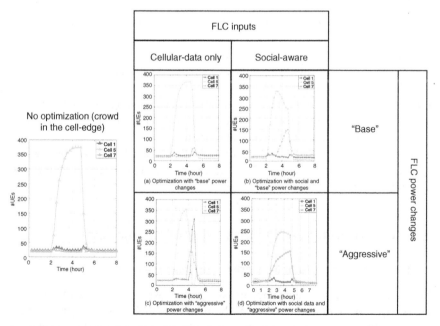

Figure 7.15 Number of users per serving cell and adjacent cells for venue located on cell edge.

the PTS controller are used: "Base", that defines slow changes in power; and "Aggressive," with higher power changes per epoch. The results show how the "base" option does not achieve enough balance in the users to avoid the overload of cell 5. For the "aggressive" option, the changes are so strong that provoke the ping pong of the users located at the social venue (in the edge of the cell) to completely move to the neighboring cell 1, leading also to imbalance.

- Social-aware optimization (right): here the complete controller of PTS+SAPTS is used, showing the capabilities of the approach to balance the network for both types of FLC values.

7.3.7 Future Directions

With the advent of 5G and beyond 5G (B5G) networks, currents network optimization practices will not be able to keep up with the network demands. In fact, as the density of wireless access points increases, resource optimization and fine tuning the various parameter settings will become exceedingly difficult to solve without the use of sophisticated AI- and ML-based methods that will make use the network-related Big Data to overcome this challenge.

In the case of B5G scenario, the challenges for network optimization can be summarized in the following points [24]:

- Resource management mechanisms based on AI and ML to exploit various diverse information.
- Network optimization to take place in an online learning manner, exploiting cutting-edge AI for intelligent and dynamic network decision making, thus achieving network automation.
- Proactive allocation of resources as opposed to responding to specific events. More specifically, by predicting the users' behaviors, traffic patterns and demands, we can improve resource allocation.
- Combining decentralized and centralized algorithms based on complexity, latency, and reliability trade-off in the context of increasingly decentralized architectures.

7.4 Location-Aware Network Failure Management

7.4.1 What Is the Cellular Network Failure Management?

Network failure management encompasses all the tasks associated with addressing service degradation. In this area, the used nomenclature have often varied among authors [25, 26]. Some common-ground terms can however be considered, as in [27]: a *problem* in the network is defined as a service *degradation* in the service provision. This is typically measure, by the reduction or increase in a specific performance-related metric, e.g. throughput and dropped calls, etc. A problem is generated due to a specific network *failure/fault* with is the *root cause* of the degradation.

Categorizations on the type of problems, depending on the affected areas of the performance, are common, typically following the classification of KPIs as defined by the standards [28]. Here, the most common categories include problems associated to *accessibility*, or the capability to receive a requested service from the network; *retainability*, or to stay connected; *integrity*, about having proper performance, such as in terms of throughput and latency; *availability*, on the likelihood of being able to connect to the network; and *mobility*, about the performance of the handover processes.

The root cause of the service degradations can include hardware failures (e.g. central processing unit (CPU) overload and antenna damage), software issues (e.g. miss-configuration) as well as context-related causes or conditions (e.g. user crowds and weather). Moreover, one or multiple problems can appear simultaneously and being caused by one or combined failures and conditions.

The aim of failure management is therefore to *detect/predict the presence of problems network and characterize them, to* diagnose their root cause, to *compensate the effects of such failures, and to* recover the network taking actions to establish it back to its original performance.

7.4.2 Why Is Localization Important in Failure Management?

In order to achieve the described objectives of failure management, both human analysis and purely automatic mechanisms generally relied on alarm analysis or performance metrics to detect the problems, identify the cause of network failures, and decide corrective actions.

However, cell-level metrics or not positioned UE traces provide a very limited perception of the network status. In contrast to this, the increasing availability of positioned UE traces implies richer data.

Here, the identified applications of location-awareness for network failure management are several:

- To enrich the available information about the performance experienced by users to identify the geographical areas affected by degradations [29].
- To allow for the geographical filtering and clustering of UEs reported information, allowing to detect isolated problems that might not be statistically significant enough to be identified by purely cell-level metrics [2].
- To better identify the cause of failures based on the association between the affected users or areas and network elements (e.g. base stations) or context ones (e.g. social events [30]), as well as to help support their prediction [31].
- To finely tune the compensation and recovery actions to the affected areas or network elements.

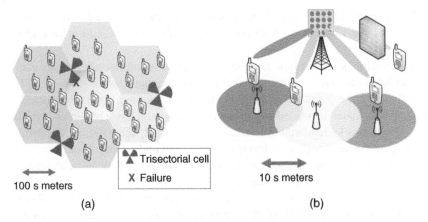

Figure 7.16 Classic (a) and 5G/6G (b) coverage scenarios representation.

- In 5G/6th generation (6G) networks, the use of highly directive antenna systems, mmWave frequencies as well as having far more dense deployments, as represented schematically in Figure 7.16 will also increase the need of positioning information to properly identified specific issues that might be directly related to the radio propagation and failures will especially affect specific spots.

The use of location information can then be utilized to support human analyses or automatic ones supported by machine learning techniques. In the following subsections the focus will be put in two approaches. Firstly, in the generation of time-series like information (contextualized indicators) and its application to diagnosis. Secondly, in the generation of image-like information called network synthetic image (NSI) for the analysis of the network status.

7.4.3 Contextualized Indicators

The purpose is to use UE location information to generate sets of indicators which combine location and cellular data. This is performed by applying specific filtering and weighting of the collected samples/measurements based on the geographical area and the position of the measurement/event. In this way, when individual UE traces are difficult to adapt to classic inference mechanisms, time-series metrics can be adopted as input for classic inference mechanisms.

7.4.3.1 Contextualized Indicators
In this scenario, contextualized indicators have been proposed as a method to combine location and network measurements in a time-series manner [32].

In brief, the approach is based on the concept of sample weights [33]. The metrics (e.g. average value, percentile value) that are gathered over the set of network measurements are statistically weighted based on the area of interest (AoI) where they have been collected. In this scenario, different sets of weights (or weight masks) can add more relevance and/ or filter one or more samples based on location. Different sets of weight functions can be applied to the same set of terminals. For example, different weights may be applied specifically to terminals using video, whose distance to the cell is less than 5 meters and served by a specific cell. This allows to give more importance to some samples than others, as well as filtering them completely depending on the AoIs and any other context variable.

Hence, the values of the generated metrics will be directly dependent on the assigned weights. In the case in which binary weights are applied to the samples, the new subsets of data will be generated by accepting or discarding a sample when they meet or not a certain geographical area criterion.

Previous approaches defined the AoIs and weights necessary to calculate the indicators manually, as well as their selection as inputs for network management purposes [32]. Conversely, we adopt a completely automated system. The objective is to define geographical areas associated to each cell coverage area, edge, and center. The areas being influenced by neighboring cells are those AoIs more likely to be affected by other cell failures, e.g. interference. Equivalently, for each cell, the area where this cell is most likely to be influencing their neighbors can also be estimated.

The estimations can be done based on propagation models specific to each scenario. A simplified approach is considered by multiplicatively weighting Voronoi tessellations, as shown in Figure 7.17.

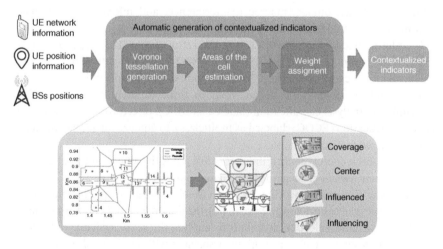

Figure 7.17 Area-based automatic generation of indicators.

This allows the calculation only based on the position of the BSs, which is an information commonly available through the OSS of the network. Hence, the cells' coverage are estimated as the areas of the tessellation. On the one hand, the cell center is defined based on circular approximations, being its radius a percentage of the distance between the BSs position and the closest point of the polygon). On the other hand, influenced and influencing areas are defined by calculating the tessellation and eliminating one cell at time.

All possible combinations of these areas (and any additional ones) are possible, e.g. cell A center influenced by cell B. From this, a huge number of indicators can be calculated, e.g. 75% RSRP percentile in the edge of a cell and average reference signal received quality (RSRQ) in the center of cell A being influenced by cell B. Overcoming this issue will be analyzed in the functionalities associated to the current application, where feature extraction techniques are expected to be applied.

7.4.3.2 Results and Conclusions

An indicative application is introduced using the contextualized indicators is introduced for diagnosis purposes.

The objective of this is to identify the presence of problems in the network and determine its root-cause. Therefore, this technique will use the contextualized indicators to determine whether there are any faulty cells and what is the underlying cause behind the failures in the cells.

The presented approach is first evaluated in an ultra-dense airport scenario with 12 LTE-radio microcells and three external macrocells with a wrap-around technique (more details on the simulator and scenario can be found in [32, 33]. Here, normal, cell outage, macrocell interference, and microcell interference failure cases are modeled for six of the cells.

In order to classify the network status, the set of classic [1, 3, 32] and contextualized indicators based on the RSRP and RSRQ measurements of the UEs are calculated for each cell, reaching 278 as the number of available indicators.

These indicators are used as an input to three different classifiers: kNN, discriminant analysis classification (labeled as DISC), or error/correcting output codes classification (ECOC) (labeled as multiclass), to analyze the impact of using contextualized metrics.

The use of the contextualized indicators has a positive effect on the diagnosis, although their use affects negatively the computational costs.

In this way, feature engineering (FE) such as selection and extraction have been used with the goal of enriching and reducing the dimensionality of has input data.

On the one hand, selection mechanisms analyze the set of available indicators and provide a set of weights or ranking associated with the level of relevance of each of them in the further classifications processes. Human experts classically performed this process, based on their knowledge on failure cases. Nonetheless, these mechanisms allow automatically establishing which indicator features can contribute the most to the system.

On the other hand, based on the statistical analysis and combination of original features, extraction mechanisms generate new "synthetic" features significantly richer in information than the original ones, triggering a more effective set of indicators which provide better information to the diagnosis mechanisms.

Selection and extraction can both be applied separately or sequentially. In this scope, three configurations are applied over the Fusion dataset. Firstly, "Fusion NCA" only makes use of the neighborhood component analysis (NCA) supervised technique for selection. Then, "Fusion NCA + PCA" adds a stage of unsupervised principal component analysis (PCA) extraction after the previous selection technique (neighbourhood component analysis (NCA)). Finally, "Fusion NCA + PPCA" performs Probabilistic PCA (PPCA) unsupervised extraction, which is regarded as an improvement over PCA.

Alternatively, the impact of reducing the dimensionality of the Fusion set without using automatic techniques is assessed via the "Manual" approach. In this, a troubleshooting expert selects a total of 24 metrics from the set of all the available indicators (both classical and contextualized). This expert follows the criteria where the chosen metrics are those that are considered most likely to clearly identify each class.

Thus, the FE techniques and classification process are repeated 100 times for each of the possible configurations to obtain statistical significance. The samples are randomly divided in each iteration, while using the same data for all configurations. Here, F1 score is used to represent the performance of the classification. This popular metric allows evaluating multi-class classifiers in terms of true and false positives (TP and FP) as well as false negatives (FN) with values up to 1, based on the expression $F1score = TP/(TP + 1/2(FP + FN))$.

In this way, the arithmetic means of all F1 score values obtained class-wise (denominated as the macro-averaged F1 score) provides a key figure of merit to compare different indicators, FE, and classifier configurations. Therefore, Figure 7.18 shows the results of the macro F1 score, where each boxplot is generated from the 100 iterations performed for each configuration.

Meanwhile, the execution times boxplots that are represented in the figure are based on 10,000 executions (100 for each of the 100 iterations indicated before) of the inference block for each configuration in a "high end" personal computer (6th Gen Intel Core i7-6700 (Quad Core 3.4 GHz, 4.0 GHz), RAM 16 GB 2400 MHz DDR4).

The fact of using a joint set of Fusion NoFE (No Feature Engineering) improves the weighted F1 score of the models in comparison with Classic indicators, since it can be also seen that this improvement is greater with discriminant analysis

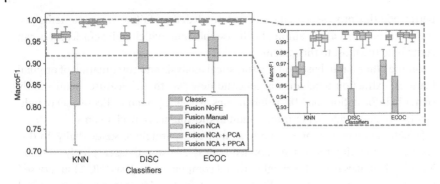

Figure 7.18 Macro-averaged F1 score [34].

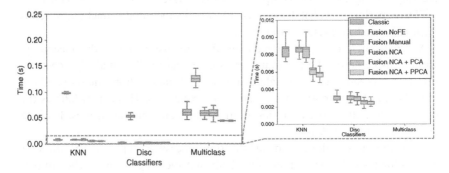

Figure 7.19 Diagnosis time assessment [34].

classification (DISC) classifier, as depicted in Figure 7.18. Nonetheless, in Figure 7.19, the increase of computational cost is seen, which this enhancement can carry out. Thus, this shows how the presence of a massive amount of indicators can contribute to improving the diagnosis, if it will also require more time. Moreover, for Fusion Manual it is assessed how, contrary to the reasoning of the expert, the manual selection leads to highly degraded F1 score.

Nonetheless, the advantage of using FE techniques mostly lies in the models' execution time. The fact of applying them allows obtaining a huge reduction in terms of time for the model to predict, which is even lower than that obtained with the other configurations. In this way, taking both F1 score and time execution into account, NCA coupled with PPCA achieve the best trade-off, which allow for positioning it has the best technique. In this way, it has been demonstrated the capabilities that might have context-awareness to improve inference in cellular networks, and the fact that using FE techniques allows for relieving the extra computational cost that they might have.

7.4.4 Location-Based Deep Learning Techniques for Network Analysis

Radio frequency maps are one of the tools traditionally available during the planning phase for decision-making, most often in the form of software simulations. Drive Tests drive tests (DT) or more recently MDT, which collect real field data during a measurement campaign carried out on a specific date in order to verify the service coverage conditions of the network, are also common during the deployment phase.

However, despite all these efforts, when the network is finally in operation, radio coverage conditions fluctuate very dynamically. This leads to a number of failures such as coverage gaps, lack of power, lack of network access or backhaul, or the sleeper cell phenomenon.

The increasing ease of access to user location data can provide tremendous advantage in determining the radio conditions at a given time and place. Depending on the accuracy of the method, it can even be assumed that the conditions at a site will be similar to points in the vicinity.

On the other hand, the refinement of deep learning (DL) techniques in the field of image processing is reaching unprecedented levels. It is not uncommon today to find applications in fields as diverse as medicine in the automatic analysis of tomographic tests for an accurate diagnosis of disease or the detection of the state of a crop in agriculture. It is in this sense that we propose to provide images of the state of a RAN in order to take advantage of the power and precision of automatic image classifiers in their management.

7.4.4.1 Synthetic mages and Deep-Learning Classification

To obtain these images, we start from the localized user traces called localized measurement (LM), which can contain different parameters of signal power, interference, throughput, among others. To be representative of the network state, they need to be collected periodically and systematically from one or several users moving around a region. Also, the accuracy of the location obtained will determine the portion of space represented by it. Once the LMs are grouped within a spatio-temporal frame, they are transformed into NSIs. The process is summarized in the following steps (see Figure 7.20):

- First, a coverage area that wants to be represented is identified; this can encompass several base stations. Subsequently, a time frame is determined that is considered relevant to the management problem to be solved; during this time, sufficient samples must be collected to cover the entire extension of the region to be represented. The NSI, being digital images, are composed of a number of pixels that represent a real square area depending on the scale (pixel size). The larger the scale, the smaller the pixel size and the more pixels make up the NSI. This is a decisive parameter, as will be seen below.

- In a second phase, the LMs within a pixel have to be grouped together in what is known as the coding stage. For this purpose, the parameter to be represented (e.g. signal-to-noise ratio (SNR)), the range of encoded values and the color assigned to represent this parameter (**RGB! (RGB!)**, **YUV! (YUV!)** or similar scheme) must be established. Thus, up to three LM parameters could be visually combined in an RGB color image, either directly or by filtering or biasing in a way that improves the representation. The latter could be achieved by applying feature engineering as a preliminary step.

- The number of LMs needed to assemble an NSI will depend on the number of monitored users and the sampling period. The pixel size is also determined by the margin of error of the positioning method. Thus, it is quite possible that a pixel contains several LMs the larger the epoch of time covered by the NSI, the higher the probability. To combine these LMs, it is proposed to use a statistic such as mean, mode, or percentiles depending on what is more relevant to the problem to be solved. A larger number of samples in the same time period increase the reliability of the NSI created, although the computational capacity required to generate it increases. Thus, the number of samples per pixel will be determined both by its size or scale and by the time epoch. With the sampling period being fixed, the time epoch that collects a sufficient number of samples to constitute an NSI can differ over time.

7.4.4.2 Results and Conclusions

As a proof of concept of the use of NSIs for the use of DL image classifiers in automated network management, a simulated dataset of an ultra-dense indoor airport environment has been used [32, 33]. In this scenario, three faults have been simulated (macro–micro interference, micro–macro interference, and cell power interruption) resulting in four states. The aim is to perform an automatic fault diagnosis based on the maps or NSIs generated by means of a supervised ML method.

Measurements have been obtained every 100 ms for 600 seconds; from these LMs of several users moving through this scenario, NSIs with different numbers of samples are created. Although the accuracy of the localization is total, different pixel sizes are proposed to simulate scenarios of greater or lesser margin of error, in this case 1, 5, and 10 m. For each of these scales, a minimum number of samples have been determined to cover the entire area covered, being 50 k samples for the 5 m option, to which the 100 k samples option is added to check whether doubling this value improves the results.

The support vector machine (SVM) method mentioned in the state of the art is established as baseline from radio frequency (RF) maps based only on RSRP with a maximum accuracy obtained below 0.9 [34]. In contrast, the use of acNSI containing 3 radio KPIs per LM encoded without bias in RGB colors and the

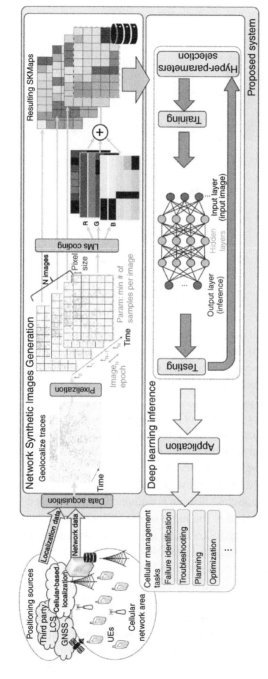

Figure 7.20 Proposed framework for NSI use and application.

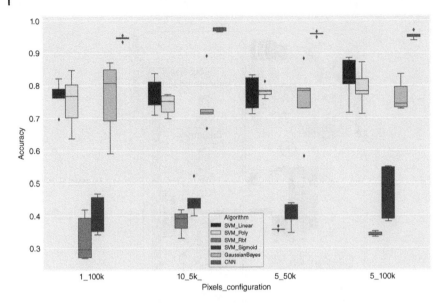

Figure 7.21 Performance of the image classification methods used.

application of a general deep learning classification mechanism of convolutional neural network (CNN)) images are proposed to show the potential improvement of the proposed framework. In order to obtain as complete an overview as possible, different choices of C, gamma, kernel hyperparameters are compared for each NSI grouping per scale.

Figure 7.21 presents a summary of the results obtained on the SVM method that serves as a baseline. According to these results, the best option in training is the use of a polynomial kernel with a scale of 10 m and only 5k samples per NSI for which a value exceeding 0.9 is achieved.

Once the best configuration for the creation of NSIs has been tested, a generic CNN algorithm for image classification is evaluated. For this purpose, a four-layer network has been designed at the input of the NSI with a 7×7 consolidation, a subsequent 2×2 clustering followed by a 3×3 convolution, and ending with a 2×2 clustering stage resulting in 4 possible outputs (states).

The results obtained for each set of NSI can be seen in Figure 7.22. In contrast to SVM, CNN achieves a much improved performance, exceeding the value of 0.9 for all options except for 5m 100 ks, reaching 92.4% in the best case.

This work has shown how the creation of NSIs can be used for various network management applications, having demonstrated a supervised network self-diagnosis method by means of an example. The creation of these images, which gives rise to subsequent refinement depending on the application for which

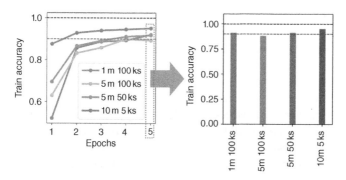

Figure 7.22 Performance of the proposed CNN approach.

it is to be used, allows the power of image classification algorithms to be harnessed simply by starting from traces positioned with a not very high precision. The effectiveness of using this approach has been demonstrated in automatic diagnosis compared to more advanced methods using RF heat maps with SVM.

References

1 S. Fortes, A. Aguilar-García, R. Barco, F. B. Barba, J. A. Fernández-luque, and A. Fernández-Durán. Management architecture for location-aware self-organizing LTE/LTE-a small cell networks. *IEEE Communications Magazine*, 53(1):294–302, 2015.

2 S. Fortes, A. Aguilar-Garcia, J. A. Fernandez-Luque, A. Garrido, and R. Barco. Context-aware self-healing: User equipment as the main source of information for small-cell indoor networks. *IEEE Transactions on Vehicular Technology*, 11(1):76–85, 2016.

3 A. Aguilar-Garcia, S. Fortes, A. F. Duran, and R. Barco. Context-aware self-optimization: Evolution based on the use case of load balancing in small-cell networks. *IEEE Vehicular Technology Magazine*, 11(1):86–95, 2016.

4 D. Palacios, S. Fortes, I. de-la Bandera, and R. Barco. Self-healing framework for next-generation networks through dimensionality reduction. *IEEE Communications Magazine*, 56(7):170–176, 2018 https://doi.org/10.1109/MCOM.2018.1700556.

5 L. Chiaraviglio, A. S. Cacciapuoti, G. Di Martino, M. Fiore, M. Montesano, D. Trucchi, and N. B. Melazzi. Planning 5G networks under EMF constraints: State of the art and vision. *IEEE Access*, 6:51021–51037, 2018.

6 L. Chiaraviglio, S. Rossetti, S. Saida, S. Bartoletti, and N. Blefari-Melazzi. "Pencil beamforming increases human exposure to electromagnetic fields": True or false? *IEEE Access*, 9:25158–25171, 2021.

7 A. Albanese, V. Sciancalepore, A. Banchs, and X. Costa-Perez. LOKO: Localization-aware roll-out planning for future mobile networks. *IEEE Transactions on Mobile Computing*, 2022. https://ieeexplore.ieee.org/document/9759997.

8 L. Chiaraviglio, A. Elzanaty, and M.-S. Alouini. Health risks associated with 5G exposure: A view from the communications engineering perspective. *IEEE Open Journal of the Communications Society*, 2:2131–2179, 2021.

9 IEC 62232:2017. *Determination of RF Field Strength, Power Density and SAR in the Vicinity of Radiocommunication Base Stations for the Purpose of Evaluating Human Exposure*. Accessed: March 8th, 2022 [Online]. Available: https://webstore.iec.ch/publication/28673.

10 A. Margaris, I. Filippas, and K. Tsagkaris. Hybrid network-spatial clustering for optimizing 5G mobile networks. *Applied Sciences*, 12(3):1203, 2022.

11 T. Hastie, R. Tibshirani, and J. Friedman. *Unsupervised Learning*, pages 485–585. Springer New York, New York, NY, 2009. ISBN 978-0-387-84858-7. doi: https://doi.org/10.1007/978-0-387-84858-714.

12 L. McInnes, J. Healy, and S. Astels. hdbscan: Hierarchical density based clustering. *Journal of Open Source Software*, 2(11):205, 2017 https://doi.org/10.21105/joss.00205.

13 N. Cressie and G. Johannesson. Fixed rank kriging for very large spatial data sets. *Journal of the Royal Statistical Society: Series B (Statistical Methodology)*, 70(1):209–226, 2008.

14 I. Hadj-Kacem, H. Braham, and S. B. Jemaa. SINR and rate distributions for downlink cellular networks. *IEEE Transactions on Wireless Communications*, 19(7):4604–4616 https://doi.org/10.1109/TWC.2020.2985681.

15 H. Braham, S. B. Jemaa, and G. Fort. Fixed rank Kriging for cellular coverage analysis. *IEEE Transactions on Vehicular Technology*, 66(5):4212–4222, 2017.

16 I. Hadj-Kacem, S. B. Jemaa, H. Braham, and A. M. Alam. SINR prediction in presence of correlated shadowing in cellular networks. *IEEE Transactions on Wireless Communications*, 21(10):8744–8756, 2022 https://doi.org/10.1109/TWC.2022.3169326.

17 V. S. Abhayawardhana, I. J. Wassell, D. Crosby, M. P. Sellars, and M. G. Brown. Comparison of empirical propagation path loss models for fixed wireless access systems. In *Proceedings of IEEE Vehicular Technology Conference (VTC)*, pages 73–77, Stockholm, Sweden, May 2015.

18 S. Saunders. *Antennas and Propagation for Wireless Communication Systems*. John Wiley & Sons, 1999.

19 M. Gudmundson. Correlation model for shadow fading in mobile radio systems. *Electronics Letters*, 27(23):2145–2146, 1991.

20 A. M. Alam, S. B. Jemaa, and T. Romary. Performance evaluation of covariance tapering for coverage mapping. In *IEEE 87th Vehicular Technology Conference (VTC Spring)*, pages 1–5, Porto, Portugal, 2018. IEEE.

21 N. Cressie. *Statistics for Spatial Data*. John Wiley & Sons, 1993.

22 L. Cressie. The sum of lognormal probability distributions in scatter transmission systems. *IRE Transactions on Communications Systems*, 8(1):57–67, 1960.

23 R. Torres, S. Fortes, E. Baena, and R. Barco. Social-aware load balancing system for crowds in cellular networks. *IEEE Access*, 9:107812–107823, 2021 https://doi.org/10.1109/ACCESS.2021.3100459.

24 C.-X. Wang, M. Di Renzo, S. Stańczak, S. Wang, and E. G. Larsson. Artificial intelligence enabled wireless networking for 5G and beyond: Recent advances and future challenges, 2020. URL https://arxiv.org/abs/2001.08159.

25 R. Barco, P. Lazaro, and P. Munoz. A unified framework for self-healing in wireless networks. *IEEE Communications Magazine*, 50(12):134–142, 2012.

26 S. Hämäläinen, H. Sanneck, and C. Sartori. *LTE Self-Organising Networks (SON): Network Management Automation for Operational Efficiency*. John Wiley & Sons, 2011. ISBN 9781119963028.

27 S. Fortes Rodríguez. *Context-Aware Self-Healing for Small Cell Networks*. PhD thesis, 2017.

28 TS 32.451. 3rd Generation Partnership Project (3GPP), Telecommunication management; Key Performance Indicators (KPI) for Evolved Universal Terrestrial Radio Access Network (E-UTRAN); Requirements (Release 17), April Release 17, TS 32.451 V17.0.0 (2022-04), 3GPP, 2022.

29 A. Gómez-Andrades, P. Muñoz, I. Serrano, and R. Barco. Automatic root cause analysis for LTE networks based on unsupervised techniques. *IEEE Transactions on Vehicular Technology*, 65(4):2369–2386, 2016 https://doi.org/10.1109/TVT.2015.2431742.

30 S. Fortes, D. Palacios, I. Serrano, and R. Barco. Applying social event data for the management of cellular networks. *IEEE Communications Magazine*, 56(11):36–43, 2018.

31 J. Villegas, E. Baena, S. Fortes, and R. Barco. Social-aware forecasting for cellular networks metrics. *IEEE Communications Letters*, 25(6):1931–1934, 2021.

32 S. Fortes, R. Barco, A. Aguilar-García, and P. Muñoz. Contextualized indicators for online failure diagnosis in cellular networks. *Computer Networks*, 82:96–113, 2015. Robust and Fault-Tolerant Communication Networks.

33 S. Fortes, R. Barco, and A. Aguilar-Garcia. Location-based distributed sleeping cell detection and root cause analysis for 5G ultra-dense networks. *EURASIP Journal on Wireless Communications and Networking*, 2016(1):149, 2016.

34 S. Fortes, C. Baena, J. Villegas, E. Baena, M. Z. Asghar, and R. Barco. Location-awareness for failure management in cellular networks: An integrated approach. *Sensors*, 21(4):1501, 2021.

Part III

Architectural Aspects for Localization and Analytics

8

Location-Based Analytics as a Service

Athina Ropodi[1], Giacomo Bernini[2], Aristotelis Margaris[1] and Kostas Tsagkaris[1]

[1] *Incelligent P.C., Athens, Greece*
[2] *Nextworks, Pisa, Italy*

8.1 Motivation for a Dedicated Platform

Chapters 1 (Section 1.2), 6, and 7 describe in detail not only the variety of the different applications, but also reveal the methodologies and the various levels of complexity of the different approaches for the definition and extraction of location-based analytics. Therefore, implementing algorithms and techniques as the ones described in Chapters 3, 6, and 7 requires a platform that will assist to this end-to-end journey for a practical exploitation of location-based analytics.

Such a platform should not only enable highly advanced and accurate positioning mechanisms, but also allow them to be fed in a seamless way to a variety of one or more analytics functions ranging from simple statistics and spatio-temporal analytics to highly complex machine learning (ML) and artificial intelligence (AI)-based mechanisms. The latter shall produce analytics results through space and time to be offered as a service to 3rd party application providers for various verticals or to the network owners themselves for the network management purposes. All these should be – at all times – subject to privacy and security aspects, adhering to rules and regulations, so as to allow for the protection of personal data and authorized access to all data and platform services.

In addition to the above, this platform by design shall be integrated with the 5G architecture and – where applicable – aligned with the concepts and technologies identified and focused on the standards specifications, studies, and technical reports available, whether related to location-based services, data analytics functions, and in general data automation concepts.

To achieve all the above objectives in a successful manner, the platform should rely on appropriate design principles that form a de-facto standard as they have

Positioning and Location-based Analytics in 5G and Beyond, First Edition.
Edited by Stefania Bartoletti and Nicola Blefari Melazzi.

Table 8.1 List of acronyms.

Acronym	Definition
3GPP	3rd Generation Partnership Project
5G	Fifth-generation
AI	Artificial intelligence
AoA	Angle-of-arrival
API	Application programming interface
BLER	Block error rate
CSV	Comma-separated value
DAG	Directed acyclic graph
ETSI	European Telecommunications Standards Institute
HDFS	Hadoop distributed file system
HTTP	Hypertext transfer protocol
IoT	Internet-of-things
ICT	Information and communication technology
JSON	JavaScript object notation
KPI	Key performance indicator
LMFs	Location management functions
MANO	Management and orchestration
MBSFN	Multicast-broadcast single-frequency network
MCH	Multicast channel
MDT	Minimization of drive tests
ML	Machine learning
NFV	Network function virtualization
OSM	Order and service manager
SaaS	Software as a service
SQL	Structured query language
ToA	Time-of-arrival
UE	User equipment
VM	Virtual machine
VNF	Virtual network function
REST	Representational state transfer
RSRP	Reference signal received power

Table 8.1 (Continued)

Acronym	Definition
RSRQ	Reference signal received quality
RSSI	Received signal strength indicator
SNMP	Simple network management protocol
URL	Uniform resource locator
XML	Extensible markup language

been incorporated into modern architecture design and fit the requirements set on the platform in question. This unique blend of platform capabilities catering to the end-to-end solution required a new type of platform that was a subject of the LOCUS project and specified in related deliverables [1]. These principles are described in Section 8.2. Table 8.1 lists the acronyms used in this chapter.

8.2 Principles

The main platform principles and design choices for a dedicated platform that accommodates all these different applications are as follows and will be described in detail in Section 8.2.1–8.2.10 (see also Figure 8.1):

- Designing a microservice-based architecture
- Exploiting software containerization technology
- Designing an AI-aware platform
- Allowing the abstraction of the computational optimization processes for the application developers
- Understanding dependencies and linking multiple blocks in an automated manner
- Utilizing a mixed Kappa and Lambda data architecture
- Targeting end-to-end low latency communication
- Decoupling API access from the actual processing
- Offering dynamic resource allocation for computation purposes
- Decoupling security aspects and services.

8.2.1 Microservice Architectural Approach

This architectural approach refers to developing an application as a complex combination of smaller, self-contained services, known as microservices. The latter can run their processes independently and communicate with other services through

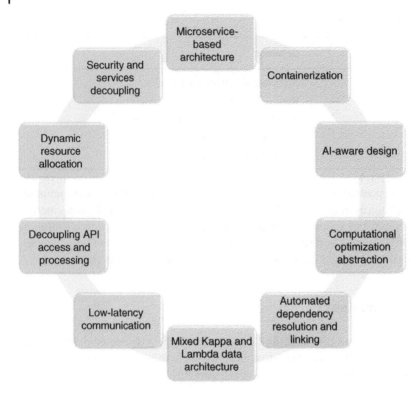

Figure 8.1 Main platform architecture principles for enabling location-based services.

predefined application programming interfaces (APIs) and messaging channels. It differs significantly from the so-called "monolithic" approach, a more traditional way of designing and building applications. An application is considered monolithic if, although consisting of multiple functions, it is developed as a single unit which cannot be divided in smaller independent sub-units and thus is deployed in a single running environment. So, despite including different functions, databases, and interfaces, all of them are unified and interconnected as a single application.

While monolithic applications are easier to deploy as a single executable file, easy to duplicate and various capabilities are tightly integrated, they lack the agility offered though the microservice approach. In fact, every change and update in the code requires a full redeployment and, depending on the size of the application, the start-up time may be significantly impacted. Furthermore, any changes, updates, and customizations can be performed and tested only with full-scale redeployment of the application. Lastly, given the variable resources requirements of the different modules/functionalities that are part of a monolithic application, scaling such an application can be ineffective and challenging.

On the other hand, the microservice-based approach – despite adding a level of complexity to achieve this modularization – allows for the highest possible degree of reusability of the microservices, agility in deployment, testability, and scalability [2]. In fact, runtime updates and upgrades are possible since microservices can be updated, deployed, restarted, and scaled in a more seamless way. Furthermore, because of their modularity, each atomic/independent function can be part of a composition of such functions, basically chaining a series of such services to provide the solution needed. This model is also fully aligned with the cloud-native approach, which aims at building and running applications, i.e. exploiting the advantages of the cloud computing concept, by caring more about making applications and deploying them, as opposed to where. This way, focus is placed at creating and operating cloud native services that automate and integrate the concept of microservices.

Based on the above, a platform whether in support of positioning, network management, or other location-based vertical applications would be highly suited to the microservice paradigm. Thus, by adopting such an approach, the platform will support the agile deployments of highly interchangeable series of functions, if needed with the 5G Core components, tailored to fit the large variety of use cases required. For instance, such a composition of microservices would allow for data collection and storage, a specific application logic, monitoring and analytics, complex ML implementations, and so on.

8.2.2 Software Containerization

Software containerization enables the microservice-based approach described previously. It refers to a technology that allows applications to run on their own runtime environment. By enabling the isolation of a software component, we allow for the avoidance of conflicts that may occur in coexisting functionalities/services and remove complexities related to installation requirements that otherwise could potentially cause serious problems in bare metal installations.

Therefore, both the various analytics services as well as the platforms systems blocks should be ideally deployed as software containers. This way – not only is the microservice approach enabled – but also installation complexities are reduced, and thus any dependencies are handled with simple commands and/or metadata. This approach far surpasses the virtual machine-based approach of silo deployments [2].

8.2.3 Mixed Kappa and Lambda Data Lake Approach

An important aspect of a platform that utilizes network and other data so as to offer location-based services is how data are handled. There is no doubt that, since we are focusing on the variety of data coming from the 5G network and other sources, we can expect data of high volume and velocity. Furthermore, these data will be an integral part of the analytics, as they will be processed for various analytics calculations and as input for ML and AI models. Since many applications have low latency requirements as well as batch or streaming data processing, as described in Chapters 1 (Section 1.2), 6, and 7, the data lakes architecture choices for the platform should allow for read, write, and transformation operations of high performance among others. A lot can be said on how a data lake approach is chosen based on the specific requirements [3]. Kappa [4] and Lambda [5] data architectures are widely used, the first more relevant with the streaming aspects of the solution, while the latter offers a batch processing layer that is also important for the analytics, ML, and AI solutions.

Since the requirements may vary depending on the application, a two-sources-of-truth approach, i.e. a mixed Kappa and Lambda data architecture would be a good approach. This can be done through the design of system blocks that will ensure there is interoperability between a given persistence module to a message queue system and vice-versa, so as to easily reuse data from different sources and go from batch to stream processing, thus supporting high-level analytics services and use cases.

8.2.4 Designing an ML- and AI-Aware Solution

The analytics services provided by the platform would require the application of various ML- and AI-based operations, i.e. implementing both unsupervised (e.g. clustering) and supervised – classification and regression tasks – along with data pre-processing and post-processing transformations. In this context, the development of models – model training and hyperparameter tuning – performance monitoring and optimization processes, and the overall lifecycle management of such models should be part of the platform design. This way, models can be deployed as analytics service blocks.

8.2.5 Abstracting Computation Optimization Processes

The computation processes are expected to include complex formulas and ML mechanisms, as well as large amounts of streaming data. For this reason, hardware optimization processes can improve performance, but they should not add a level of complexity for the developer of an analytics service. The system should obscure such details and separate the application/analytics aspects. This is done through a variety of modern libraries and frameworks, such as Apache Spark [6], Seldon [7] and others. This way, the developer does not have to be knowledgeable with respect to the optimization aspects in order to develop any processing or ML-related task, and thus this can help developers to offer solutions with less effort.

8.2.6 Automating Dependency Resolution and Linking

As each location-based analytics service is expected to be composed of a chain of smaller building blocks, thus allowing for the reusage of specific functions, the system should somehow have knowledge of this linking and the various internal dependencies, so as to be resolved and orchestrated accordingly. For instance different services for network optimization and flow monitoring could reuse the same data collection function from a 5G network and a positioning function given a specific area, then these data/results are further utilized for enabling the services in question.

Internal dependencies are declared to the system using meta-data so that dependency resolution and better resource utilization can be done via an orchestration entity. Services can be either directly linked via their internal interfaces or they can exchange data and results through the persistence and message queue blocks.

8.2.7 Achieving Low Latency End-to-End

Since many of the enabled services require low or extremely low latency requirements, e.g. detection of possible vehicles collisions, vulnerable road user applications, etc., the selected technologies should allow for this. This requirement involves various systems blocks, each contributing at a different degree to the end-to end latency, i.e. blocks related to the data ingestion phase and related processes such as data cleaning and anonymization, inter-components data exchanges, the data processing and ML phase, but also the system components related to the analytics delivery mechanisms to 3rd parties. Thus, aiming – at least for specific analytics – to be developed and delivered in a streaming context will allow for the lowest possible end-to-end delay toward delivery.

8.2.8 Decoupling Processing and API Access

To avoid extreme resource usage leading to overall unstable system operations, some orchestration of the tasks given the various service requests should be considered. If this is not the case, the system receiving directly requests may fail to secure the appropriate resources due to unavailability and/or incorrect task prioritizations scheduling, thus leading to unreliable overall performance.

For these reasons, the analytics service request could be propagated to a dedicated entity which will then orchestrate the execution of each task ensuring resource availability, execution order, and deliver from an API the analytics service output by interfacing with the data repository designated to this service. The API itself will provide contextual information of the status of the request (e.g. if it is ready or under processing, etc.).

8.2.9 Offering Dynamic Resource Allocation

Resource allocation is essential for the smooth operation of the platform. As different services requests are received and various internal operations are under way, scalability and optimal placement of the resources are essential and impossible to perform manually in an optimal way.

The different variations in demands and applications should be handled by a virtualization layer and the dynamic service discovery engines residing within the system. Through this layer, analytics demands and resource availability can be balanced through intelligent matching in an automated way.

8.2.10 Decoupling Services and Security

As the need for security is essential for any system, it is even more so for a system that incorporates telecommunications-related data (network and customer). For this reason, and based on the microservice architecture which adds a level of complexity on the enforcement of security, security functions should apply on various stages of the platform. This can add time and effort for developers.

As a minimum, therefore, the system shall have dedicated system blocks for the enforcement of the security requirements with respect to the external access. However, internal microservices can use a Single Sign-On interface to replicate user information and utilize it internally, allowing for developer to not focus on the security aspects, but the service under development itself since their security features will be limited. This way, services are easier to both develop and maintain.

8.3 Platform System Overview

In order to offer location-based analytics as a service through a dedicated platform, the latter should incorporate some essential system blocks that will enable such services. As mentioned in Section 8.2, each block is an individual system or a set of such systems that exist within their own container on top of the virtualization layer. Furthermore, they can be divided in categories/types based on the related functionalities they offer and on best practices to ensure flexibility in development and deployment. A high-level system block view of this architecture is shown in Figure 8.2.

Specifically, the blocks include a layer related to the virtualization management and infrastructure, platform control components, core components (i.e. related to the analytics, persistence and message queue blocks, and other services), and the API layer components. Also, it is worth noting that – while not part of the actual platform – two different categories are presented, as they are tightly connected

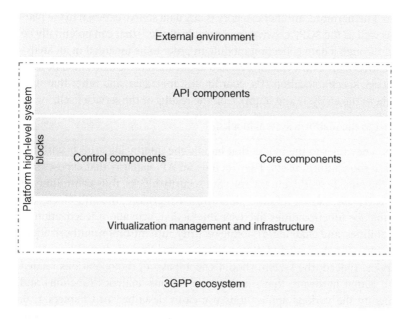

Figure 8.2 System blocks' categorization for a location-based Analytics as a Service platform.

with the platform; the 3GPP ecosystem and the external environment blocks. In more detail:

3GPP ecosystem blocks: By this term, we refer to related 3rd Generation Partnership Project (3GPP) infrastructure. It is no surprise that a tight integration with the 3GPP infrastructure is expected and should be targeted from the design phase of the platform.

Whether for the extraction of positioning information or other analytics based on the aforementioned information, the system must have access to network data, e.g. network measurement jobs in the various network management entities systems, which will be utilized. The data feeds from this ecosystem can either refer to batch exports of information, as well as cases of streaming data. Similarly, the results of the analytics can be exploited in order to assist network management-related or 5th Generation release-related operations with respect to Internet of Things (IoT), self-driving cars, smart cities, etc., thus proving once again the 3GPP ecosystem relevance to the platform.

External environment blocks: External environment blocks are all blocks relating to various systems and stakeholders that may have active or passive interactions with the platform. These involve entities that subscribe and/or consume the actual analytics provided by the platform and any 3rd party users of the platform. Furthermore, another category is any data source external to the platform – as well as the 3GPP ecosystem exposable blocks – that can potentially be ingested through a data collection module in order to be included in an analysis. Indicatively, for a flow tracking related application these data could be map information, specific location IDs, coordinates, metadata, and other that may complement the analysis and improve on the results or the service itself.

With respect to the platform system blocks:

API blocks: They refer to the blocks that handle the interfacing aspects with the external to the platform users. They include an API gateway that exposes services to the outside world and the relevant security blocks, thus controlling all the 3rd party user access into the platform. Furthermore, the API blocks are responsible for functionalities like user interaction, activation/deactivation of API capabilities, and enabling/disabling security authentication/authorization.

Control blocks: The dedicated platform aims at providing on-demand analytics as a service. Indeed, the system should encompass all the operations related with the actual hardware that supports the various analytics functions and subsequently the various applications/use cases described in Chapters 1, 6, and 7. These function lifecycles are varied with respect to their different dependencies, data employed, ML/AI operations. The control blocks of the platform facilitate operations like the replication/instantiation of requested resources, the management multiple simultaneous requests through re-parametrization of batch computations to serve multiple simultaneous requests. In fact, the control blocks bridge that API access request with the virtualization technologies of the platform enabling software as a service (SaaS) system capabilities.

Virtualization management and infrastructure blocks: This category includes blocks related to the management, orchestration, and virtualization aspects of the platform. Their aim is to have virtual functions and services deployed over a common virtualization platform, and in this way abstract the actual infrastructure of the platform, whether for edge or core computing locations. Adding to this, blocks dedicated to the management and orchestration will allow for the seamless deployment of location-based analytics services and functions as virtual services/functions, through the provision of resources and orchestration logic, e.g. for when to instantiate a function and more.

Core blocks: They involve a highly varied collection of blocks, including:

- Persistence and message queue blocks: Related to data persistence, data streaming, and messaging. They are an integral part of the platform, as the handling of data – whether from outside data sources or within the platform to be used among the different components/functions – is of high importance. They should accommodate a dual stack of streaming and batch storage in order to fit all the functional requirements discussed in the architecture principles in Section 8.2, but also the interoperability between the two data lake modes, translating the persistence (in the form of structured query language [SQL]-type tables) into messages and topics (for the message queue) and vice-versa.

- Other core service blocks: They can be split in (i) the analytics and positioning-related blocks, responsible for producing intermediate or finalized results for the positioning, analytics for the platform users, to be exposed at the acAPI components' layer; (ii) the data collection module, in charge of collecting information either from the 3GPP ecosystem data sources or from external 3rd party data sources whether in batches or real-time; (iii) the anonymization module that provides different services/functions for the abstraction of private/sensitive information through data manipulation and transformations.

In Section 8.4, the platform system blocks are described in more detail.

8.4 Platform System Blocks Description

8.4.1 API Blocks

A suggested division in different blocks is presented in Figure 8.3. They refer to blocks related to user interaction, profile, activation/deactivation of API capabilities, etc., the API gateway and security-related blocks that control 3rd party user access into the platform. In more detail:

- Service subscription module: This component handles user registration in the available services and triggering all the necessary processes for the selected service. It provides information about the platform analytics services in the form of meta-data included in the API catalogue. Furthermore, it is connected to (i) the access control module, which ensures the relevant permissions for platform users; (ii) the service discovery module in order to ensure the readiness of the instantiated service.

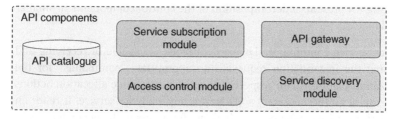

Figure 8.3 Platform's API block components.

- API catalogue: It allows for the explicit description of all the available services for users accessing the platform. The information is provided in the form of metadata and is related to the service subscription module and other blocks that require the sharing of this information. For new services that are onboarded and/or changed, these metadata are updated accordingly.
- Access control module: This is highly connected with the need for security when providing analytics to other parties. Access rights should be specified and users should be allowed to the access the various analytics based on these rights. For this reason, such a block would enable user generation, administration, and association with the aforementioned rights, as well as enforce the credential verification or key-based access on all resources.
- Service discovery module: An important aspect of the platform is to enable flexibility in the offered services, i.e. to activate/de-activate the various service capabilities based on demand. This module keeps track of the internal systems, and it communicates with the service subscription module to provide information to the user about the status of the requested API access, thus linking the external API and the actual status of the internal services.
- API gateway: This block is of high-importance, as it acts as the connecting link between the platform's external environment – which in this case acts as the consumer of the services – and the analytics results that lie within the platform. Through its link with the access control module, it enforces user access, while it also includes load balancing capabilities and https-based transport protocols to ensure standard internet security.

8.4.2 Control Blocks

These blocks are essential for the smooth operations within the platform and the enablement of location-based services. Given the variety of the various use cases, a large number of highly diverse functions with different hardware requirements, dependencies to other functions and/or data are expected. Providing on-demand analytics requires a system that links the requests of the API layer to the virtualization technologies and controls operations, replicating/instantiating resources, and so on. Figure 8.4 presents a decomposition of the control components in smaller functional blocks.

- Analytics service coordinator: This system block is capable of receiving and translating the instantiation requests into appropriate "commands" for the virtualization platform with respect to the needed resource allocation actions and startup of all the necessary software for any analytics service. It relies on the analytics service catalogue, which allows for access to the dependency and execution requirement descriptors of each service. Based on this information

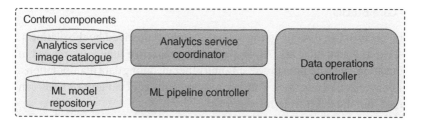

Figure 8.4 Platform's control block components.

the instantiation of the analytics service and its related functions as virtual service and virtual functions (e.g. containers) are triggered by the Management, Orchestration, and Virtualization Infrastructure blocks, and service execution and runtime configuration is orchestrated.

- Analytics service catalogue: This block catalogues all service descriptors. The latter describe the services' capabilities and related execution requirements (including reference to virtual resource requirements descriptors). It is linked to the analytics service coordinator, providing the necessary information for the control operations mentioned above.
- Data operations controller: In general, a data operations controller fits the Lambda data lake architecture [5], where a central dispatcher coordinates the various big data operations that manipulate data from the various locations within the data lake. In detail, this block receives configuration from the API access layer or internal triggers to perform processes/computations for the analytics services. It further communicates through a control interface and monitors the various analytics services currently running, while working in harmony with the analytics service coordinator, in order to request more replication of the same analytics to cater to the various demands. In this sense, it can be characterized as a high-level orchestrator and scheduler of the aforementioned tasks.
- ML model repository: This block is directly linked with the principle of designing an AI/ML-aware platform as mentioned in Section 8.2. In more detail, many services leverage AI/ML-models and for this reason the platform should cater to the needs of such models, i.e. aspects related to the continuous training, evaluation, and re-evaluation cycles necessary. This block allows for the support of such operations, reducing the need for more customized implementations.
- ML pipeline controller: As with ML model repository, this system block of the platform is related to the AI/ML-model capabilities. It provides a programmatic API layer between the analytics function and the management and orchestration (MANO) and virtualized infrastructure that tries to enforce embedded hyper-parameter tuning and containerized preprocessing pipelines with minimum developer interaction.

8.4.3 Core Blocks

The core blocks relate directly to the analytics services provided by the platform, they refer to the analytics, ML/AI, and relevant supportive functions, related to data acquisition, pre- and post-processing and other incoming data manipulations, e.g. for anonymization purposes. They are presented in detail in Figure 8.5 and can be grouped into the following subcategories:

Persistence and message queue: These system blocks are connected to the data lake architecture chosen for the platform. As the choice for this platform is the utilization of a dual stack of streaming and tabular storage so as to accommodate a variety of different use cases, these blocks cover the need for data movement, persistence, and messaging. In detail:

- Persistence module: This module is dedicated to data persistence. The capabilities of such a module should include the provision of an environment for storing both structured (tabular) and unstructured data. This block is not necessarily utilizing a containerized deployment, a choice of bare metal deployment is preferable to achieve the highest possible input/output speed performance. Furthermore, the choice of technologies should enable however scaling-out capabilities, i.e. the increase and/or decrease of nodes depending on needs. Lastly, it is expected to support write interfaces for raw data transformations into structured and big data query engines, i.e. for fast read operations, aggregation functions, window functions, filtering and selection.
- Message queue module: It supports the need for streaming data operations through publish–subscribe mechanisms whether serving needs within the platform (i.e. internal module intercommunication) or communicating with 3rd party data consumers that need low latency data delivery. It is important that it allows for multiple topics, formats, network protocols, as well as temporary persistence capabilities (i.e. allowing retention periods for messages). Also, due to the need for real-time data delivery high read (processing) capabilities should also be supported.
- Interoperability functionalities: While not explicitly mentioned in the figure, these should be supported with a series of interoperability functions that

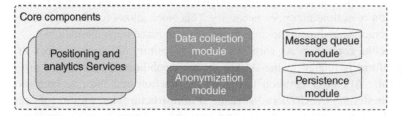

Figure 8.5 Platform's core block components.

allow translation of SQL-type tables into topics and messages of a specific topic, and similarly conversion of messages into insertion/update/delete events, thus allowing the interfacing and supporting both persistence and streaming stacks.

Positioning and analytics: They refer to the positioning and analytics functions that can be instantiated whether internally within platform to support other functions or to be exposed to outside parties (after being discovered by the service discovery module) through the API gateway. More specifically, positioning functions allow for collected 3GPP- and non-3GPP-based input data (delivered from the data collection module) to be translated into positioning information of the user equipments (UEs) involved. These service blocks are not necessarily directly provided by the API gateway to external users, but their generated positions can act as input to the other analytics functions.

The analytics functions are services that are also instantiated within the platform with the purpose of producing analytics, ranging from simple key performance indicator (KPI) calculations and statistics to high-complex AI-based results. While they can be very diverse, an obvious division of such services is the actual purpose, i.e. targeted applications for these analytics. In this context, they can be split into analytics functions catering to network management applications as in Chapter 7 or other vertical use cases, as indicatively presented in Chapter 6.

Data collection and anonymization: In order to enable the previously described analytics, specific functionalities related to the data to be utilized should be taken into consideration. Indeed, data collection capabilities would be essential in order to allow the gathering of information from diverse sources, in various formats, delivered in different ways to the platform. In addition to this, in order to take into account any privacy considerations and to allow for the sharing of analytics results that do not expose personal information in any way, an anonymization framework should be considered. In more detail:

- Data collection module: This module enables the collection of data from internal (3GPP) and other internet-based sources that may vary significantly and saving the information in the various system persistence and/or message mechanisms. It optionally can employ the anonymization functions of the anonymization module in order to transform/alter the data in a way that will guarantee no violation of privacy will occur.
- Anonymization module: It is a module dedicated in the abstraction of sensitive information, safeguarding privacy. It manipulates data so as to offer anonymization of (i) in the context of the data collection phase – also related to the ingestion phase – and (ii) during the ingestion of external information that can be anonymized before its persistence in the persistence module or the message queue.

8.4.4   Virtualization Management and Infrastructure Blocks

An infrastructure virtualization layer is necessary in order to abstract edge and core computing locations to provide a common environment where virtual functions and services can be deployed. Furthermore, by modeling localization analytics services and functions as virtual services and functions and offering management and orchestration system capabilities for the required coordination logic, location-based analytics services are enabled. The relevant blocks are mentioned briefly below (see also Figure 8.6):

- Management and orchestration: This block includes all the management and orchestration of virtual resources capabilities allowing for the automation of the services. It allows required virtual functions and services to fulfill the analytics service requirements related to instantiation, scaling, runtime update, monitoring, and termination as expressed by the platform control bocks through the dynamic deployment and operation of the required virtual resources.
- Virtualization layer: This layer provides a common access to virtual resources (network, compute, storage) in the distributed edge/core infrastructure, as required to run the various localization analytics services and functions. In this context, it will allow for the utilization of virtual resources for both static and dynamic system components. While a bare-metal allocation of virtual machines may be chosen for specific platform blocks, it should primarily use container-based virtualization.

8.5   Functional Decomposition

As described in Section 8.4, each of the above system blocks has multiple capabilities that can be translated in one or more independent chainable functions per capability. Depending on the type of functionalities the platform offers, each function can display various degrees of complexity and different parameterization aspects for their instantiation. In fact, a developer of, e.g. an analytics service, may

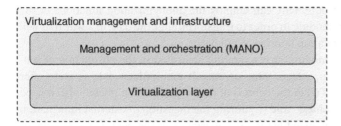

Figure 8.6   Platform's management, orchestration, and infrastructure components.

decide to include one or more functionalities in one function or offer multiple functions for the same functionality if more than one method exists.

In Sections 8.5.1–8.5.7, a description of representative functions for a system catering to location-based analytics is presented, based on the work done within the LOCUS project [1].

8.5.1   Data Collection Functions

The data collection functionalities can be offered by one or more data collection functions sufficiently parameterized to cater to the diverse cases of data sources, the types of data, the data transportation protocols, interaction mechanisms used, and their specific parameters if applicable. A data collection function is envisaged to relate to the following aspects:

Data categories and sources: More specifically, the platform is expected to collect various types of data and offer both streaming and batch ingestion capabilities. Based on the various network-related and vertical applications described in this book, these data could be roughly categorized in the following:

- Network-related data: These types of data can vary significantly. For positioning data, they include positioning-related KPIs, such as time-of-arrival (ToA), angle-of-arrival (AoA), received signal strength indicator (RSSI). Other network metadata that are also important are those related to Network management/configuration, e.g. network topology and cell metadata (e.g. position and various configuration parameters), positioning information coming from 5G core location management functions (LMFs) [8]. Other 5G network KPIs include for example monitoring and management KPIs, such as the traffic, and that are available from various sources at a cell level or an overall area under investigation (e.g. a venue, an airport, a town's center).
- Device data: These are the data that come directly from the devices, such as data from mobile's barometer, Global positioning system (GPS) data and other.
- External/other data: Similarly, any auxiliary information that can be used to offer location-based analytics can be included. For example, IoT sensor data, data from other internet/cloud sources that may assist the applications such as geometry data (e.g. areas, maps, geometry metadata for area representation of structures/roads), other multimedia data (e.g. from cameras). Even data from media feeds and social media coming from 3rd party could be part of an analysis, e.g. for people flow monitoring and crowd control applications.

From the aforementioned categories a wealth of different data sources, sufficiently parametrized in order to be ingested are involved. These include: (i) 3GPP-compliant network/non-network sources that have a description in

JavaScript Object Notation (JSON), (e)Xtensible Markup Language (XML) or other similar format, (ii) internet sources with hypertext transfer protocol (HTTP)-based or other internet protocols requiring a URL and authentication parameters and other cloud services requiring some user access information, (iii) other sources of network data that could be retrieved, for example through Simple Network Management Protocol (SNMP), and other protocols.

Batch and streaming data: The way these data should be collected is through two data transportation protocols:

- HTTP/REpresentational State Transfer (REST): Whether client (external to the platform) or server-side (i.e. platform), an HTTP/REST API is used to query and receive the data in batch form.
- Message bus: It refers to exchanges of messages on top of a message broker from streaming data applications. In this case the data collection function consumes the continuous data stream. It can work both ways, i.e. the platform consuming from an external broker, or the external source is a producer in the platform message bus.

From the above description, in both cases, we can envision the communication to be initiated from the platform or the data source or both (mixed approach). Another important aspect refers to rate of acquisition and the relevant details in connection with the above. More specifically, a periodic acquisition of data could be achieved through a polling approach where the periodicity can be diverse. On the otherhand, a one-time request/response process can help data acquisition in an asynchronous manner, while in the publish–subscribe (pub–sub) paradigm streaming data related to a specific topic are exchanged in a streaming manner.

8.5.2 Persistence and Message Queue Functions

Following the data acquisition process, the persistence of the data and the consumption of the streaming information are of essence. The related platform functions should allow for the persistence of both structured and unstructured data, the transformations for one type to another, the ability to query specific data, to create and destroy topics, to consume and publish data, as well as ensuring interoperability between the data lake and the message bus. In short, the relevant functions can be split to three major groups. The ones referring to data management/persistence, the message queue-related, and those allowing for the interoperability of the former.

The persistence module is seen as a repository for both structured, i.e. relational, in the form of rows and columns, and unstructured/semi-structured non-tabular raw data. It should also allow for ingesting real-time, batch and data streams, and store other processed data coming from internal functions, in structured and/or unstructured formats. For this reason, persistence, data manipulation, and transformation functions should be considered. Similarly, streaming data should be supported both as an alternative data provisioning for 3rd parties, but also due to the possible streaming nature of an analytics function. Thus, message exchange operations through a message broker should be considered, as well as interoperability functions converting data from the persistence module into messages and vice versa. All the aforementioned functions are included in Table 8.2.

Table 8.2 Persistence data management and message queue functions.

Function(s)	Description
Structured data persistence	This function allows the persistence of relational data multi-dimensional datasets of numerical and categorical features into the data store. structured query language (SQL)-compliant methods are used for data retrieval
Unstructured data persistence	Used to persist object-based and document-based data (e.g. blob/serialized/text) into the data store
Structured and unstructured data query	These functions allow access to (i) structured data with an SQL-compliant format query, (ii) unstructured data with a query defined based on the dataset's index, respectively
Dataset transformation	This category includes a set of functions that do the following: • Transformation of tabular-based datasets into semi-structured or unstructured forms, based on a provided mapping logic • Transformation of an unstructured object into a multidimensional structured dataset, given a mapping function and a target dataset structure • Transformation of an unstructured object of a given type into another unstructured object. The mapping logic should be provided as input
Generate/destroy topic	Used for creation/destruction of a message queue topic for a given analytics function and its operational status on demand
Produce/consume data	Allowing the production or consumption of data, given a specific topic in the internal message queue
3rd party data consume	This function delivers streaming analytics data to 3rd parties (respecting security and privacy requirements)
Generic sink into landing	Converts messages into entries in a designated landing schema
Persistent change data capture into messages	Converts batch data operations (save or update) into message queue events

8.5.3 Positioning and Analytics Functions

8.5.3.1 Positioning Functions

The positioning functions provide the necessary mechanisms for the positioning of the user equipment (UE) (or user). As presented in Part I, they can vary significantly and utilize different approaches, but the expected output is the position of the target, which should meet some criteria on the accuracy, reliability, and update rate given a specific application requirement. Their results could be directly exposed to other parties – assuming security and privacy criteria are met – or be used as input in other analytics functions. In this context, positioning functions could be categorized based on the type of data used as well as by the mechanisms to enable positioning. In more detail, they could be differentiated as follows:

- 3GPP technologies-based: By leveraging 3GPP defined radio measurements along with information of the location of access points, these functions purpose is to derive the UE position(s) meeting a defined set of quality criteria.
- Non-3GPP and hybrid technologies-based: In this case, non-3GPP data related to received signal power, time delay, angle of arrival/departure are used as input either on their own or in conjunction with 3GPP data to derive the UE position.
- Device-free localization-based: The positioning of one of more device-free targets (e.g. persons, vehicles) is done based on a passive monitoring system and through the identification and analysis of backscattered signals.

8.5.3.2 Analytics Functions

A wide variety of location-based analytics could be part of a platform that leverages positioning information and thus be exposed as a service to other applications. These location-based analytics could employ a wide range of capabilities starting from simple statistics to exploiting highly advanced ML and AI approaches. Furthermore, they can be split based on the type of service they enable or the type of methods they employ. In terms of the type of services, they can be split in Network Management – related and vertical/3rd party applications. In terms of analytics used, they can also be split roughly into the following categories:

- Descriptive analytics: i.e. statistical analysis on collected/historical data aiming at understanding what has already happened.
- Predictive analytics: i.e. predicting/assessing future possibilities based on collected information.
- Prescriptive analytics: i.e. given known information to prescribing specific actions to be taken in order to achieve one or more objectives.
- Diagnostic analytics: i.e. understanding based on the collected information what were the causes of a specific outcome.

Indicative implementations of such analytics can be found in Chapters 6 and 7 and were a subject of the LOCUS project [1] where a detailed description of such analytics functions can be found in the relevant deliverable. Table 8.3 summarizes a rough grouping of such analytics functions focusing on location-based network management and mobility/flow monitoring applications. It should be noted that these functions assume knowledge of the positioning information of all or selected targets in a given area.

It is worth mentioning that several other such functions could be part of the services offered. Furthermore, these analytics functions (i) may be decomposed into smaller functions (such as one or more preprocessing/data manipulation procedures of the positioning and other incoming data, followed by application of ML and/or AI methods, and then basic analytics/statistics on top of the resulting data), and (ii) may be combined to create a more complex service.

8.5.4 Security and Privacy Functions

Security and privacy are critical aspects of any platform, especially a platform offering access to location-based analytics. It is also important to separate security and privacy concerns from the location-based application functions, as this way all developers can focus on specific localization features using a common approach to retrieve security and privacy functions. In this context, the location security and privacy functions shall ensure all data privacy and security aspects through the APIs that handle data acquisition and sharing among the system blocks. The relevant requirements and concerns are presented in more detail in Chapter 5; however, we list here relevant functions and a short description.

8.5.4.1 Security Functions

- Authentication/authorization: This function allows for user/device authentication and authorization. For all network interfaces, it implements cryptographic techniques and is in line with the Lawful Interception Architecture defined in 3GPP [9].
- Data clustering for security purposes: This function utilizes ML-based and specifically clustering approaches for anomaly detection to achieve separability between the useful signal component features/parameters and those of the interference signal thus mitigating erroneous measurements. Ideally, its output may include a flag on possible ongoing attack, labeling of measurements, and the actual mitigated measurements.
- Data cleaning for security purposes: This function aims to detect and mitigate (noise) jamming and spoofing/meaconing attacks ensuring the reliability of the location estimates. It can use methods ranging from summary statistics to ML-based strategies.

Table 8.3 Indicative list of location-based analytics functions.

Function(s)	Description
Trajectory detection	This function produces identified trajectories for a given area and time window based on collected data. It can be agnostic to the positioning function employed given specific quality requirements apply
Point/area of interest detection	It allows for the identification of specific areas described using some geometric representation (e.g. a polygon). These areas are derived as a result of analyzing information collected, such as high UE density and others
Path/route identification	Assuming a point of origin and destination and a given time-related constraint, it produces the identified timed trajectories/routes. This function could also trigger a predictive model for future time windows for predicting the best route
Flows identification	Similarly, this function can produce an origin-destination matrix of velocity fields for a given area whether for a given time window in the past or through predictive analytics for a time window in the near future
Count identification	This function defines the number of individual targets (UEs or users) found in a given area or predict how many will be found in a given area
Mobility changes tracking	This function can support a subscription request to monitor a specific mobility-related feature (e.g. velocity, direction, and proximity) of an individual or group trajectory and receive the relevant notification when a specific condition applies
Time-to-collision	It estimates the time and likelihood of collision of moving objects
Mobility profiling	It allows for the identification of one or more user trajectory profiles from a set of given profiles, e.g. "pedestrian," "vehicle - high velocity"
KPI heatmap	This function generates a heatmap view of a given KPI, whether historical or predicted. It may also refer to the mapping a KPI to ML-derived areas or trajectories
KPI prediction	This function utilizes ML and AI to predict network-related KPIs, such as network demand and network quality metrics, in the near future
Contextualized indicator generation	This function enables the fusion of diverse time-dependent data in a selected area and time-frame. It is employed to allow for the dimensionality reduction of possible large amounts of indicators and can be utilized as input to other functions
Others, basic analytics functions	This group of functions refers to simpler analytics functions that usually complement the more complex ones and refer to location and KPI data manipulation. Such functions include correlation of geometric areas, geometric area manipulation, data filtering, metric and categorical aggregation, geo-query (i.e. identifying points to specific areas), reverse geo-coding (i.e. transforming geolocation data into areas that contain these data), etc.

8.5.4.2 Privacy Functions

- k-anonymity: This function produces data so as individual entities are indistinguishable from another k-1 individuals.
- Sanitization, obfuscation, and result aggregation: They aim at concealing sensitive data by (i) removing user information from the stored location data, (ii) blurring/perturbing the location data, and (iii) allowing the delivery of only aggregated information to 3rd parties feature queries, respectively.
- Policy definition: This function allows the management of different levels of security and privacy and definition of user/device policies to ensure this. In this context, it relates to filtering, aggregation, anonymization, and obfuscation procedures.

8.5.5 Analytics API Functions

The exposed services are a composition of one or more functions chained together based on their specific input/output data requirements so as to provide a comprehensive analytics service. The analytics API functionalities, provided by the API block, allow to query, subscribe, and activate analytics and services and functions on-demand. Furthermore, there are other capabilities supported, such as access control and dynamic service discovery. The main functionalities provided are as follows:

- User access control: It is essentially an interface between the security functions for authorization and the service subscription function, linking platform users to their respective catalogued and active analytics functions.
- Service discovery: The various (generated on-demand) analytics function instances register their availability for automated access.
- Service metadata manipulation: This function allows for the creation, update, and deletion of analytics metadata information in the API catalogue where they are stored.
- Service subscription: This function of the API block relates to the exposure of service offers through asynchronous service request/response transactions that provide access to the service descriptions in the service catalogue to dynamically activate (and deactivate) analytics services and functions.
- Analytics 3rd party consumption: Through an asynchronous service request/response transaction, the localization and analytics service consumption endpoint is delivered to external applications. The API gateway supports both HTTP REST and message bus-based approaches for the exposure of services.

8.5.6 Control Functions

To ensure that the various analytics services are onboarded and executed in an orchestrated and dynamic way, control functions need to be implemented. They

Table 8.4 List of analytics API functions.

Function(s)	Description
Analytics service instantiation/ termination	Whether triggered from a 3rd party application event (e.g. through a service subscription and activation) or from an internal pipeline, this function is responsible for activating and replicating the various analytics functions. In this manner, each function invocation results in the instantiation (or termination) of a replica of an analytics function
Service catalogue entry persist/ query	This function enables the saving of service capabilities and characteristics information (data requirements, virtual resource requirements and software images) into the analytics service catalogue. This is crucial for the activation and runtime operation of the various analytics services/functions, as the information must be up-to-date so that the analytics service coordinator can request the virtual resource allocation operations and the necessary execution logic can be applied
Analytics service configuration	This function reads the desired analytics service configuration and internal dependencies of the given service and manages the execution of the analytics functions and services, e.g. when an analytics pipeline is required among the lower-level functions. The pipeline is then executed assuming that the requirements are met
Analytics Service Health Check	The spawned services health information (e.g. running and error) must be available to the data operations controller, as the latter supports parallel and sequential subtask executions described by direct acyclic graphs (DAGs)

include (i) the data operations functions for the high-level orchestration of analytics services; (ii) the service coordination functions, which act as mediator toward the virtualization layer (and related management and orchestration functionalities); and (iii) the functions that facilitate the high performance computing, ML model lifecycle support, data pipeline execution virtualization, and coordination. Table 8.4 presents these control functions.

8.5.7 Management, Orchestration, and Virtualization Functions

As mentioned in Section 8.4, a virtualization platform provides an abstraction layer of the actual infrastructure, thus exposing a unified virtualized infrastructure for the deployment and execution of the various virtualized functions. Additionally, a management and orchestration framework allows for the automated provisioning of location-based analytics functions and their deployment, configuration, and operation on top of the virtualization platform as virtualized functions. Based on this, a number of related functions can be enumerated (See Table 8.5).

Table 8.5 List of control functions.

Function(s)	Description
Virtual resource management	It handles the creation, update, or deletion of virtual resources (incl. compute, network and storage resources) based on the characteristics and requirements given as inputs in order to execute the various positioning and analytics functions
Container management	As above, the creation, update, or deletion of containers (or PODs) so as to execute the various positioning and analytics functions as containerized functions
Virtual resource monitoring	This functions allows the consumption of information regarding the use and status of the virtual resources, so that they are retrieved and used by the MANO
Service orchestration	This function coordinates the lifecycle management actions for deploying the analytics services (modeled as interconnected and virtualized individual positioning and analytics functions) in the form of Network Function Virtualization (NFV) network services. It uses the resource orchestration function to manage the various Virtualized Network Functions (VNFs)
Resource orchestration	This function delivers lifecycle management actions (e.g. instantiate scale) for the virtualized analytics functions (modeled as VNFs) executed through interfacing with the virtual resource management functions in the virtualization platform
Service optimization	This function adjusts NFV network services at runtime in an automated manner to improve service topology, size, and deployment, considering both the actual and predicted performances
Fault management	This function allows for the automatic management of detected or predicted failures occurring in the virtualized platform or the VNF instances. This is done through the service orchestration function in order to proceed with lifecycle management operations
VNF and NFV network service catalogue operations	It includes an on-board and a query operation that stores and queries VNF Descriptor or Network Service Descriptor information. Container image repository operations As above, it includes an on-board and a query operation related to the container image repository
Virtual machine image repository operations	Similarly, allows onboarding and querying software images of positioning and analytics virtual functions deployed as virtual machines

8.6   System Workflows and Data Schema Analysis

8.6.1   System Workflows

Starting from the location-based analytics as a service platform system overview and blocks description, a set of operational workflows have to be supported and implemented to let the platform support and offer the required functionalities. Specifically, the various system blocks are required to tightly cooperate and interface each other to enable analytics service activation in the virtualized infrastructure, analytics service consumption through the API block, data collection, and execution of the specific positioning and analytics functions and services. In practice, the following system workflows are envisaged for the proposed location-based analytics as a service platform:

- Service activation system workflow
- Service consumption system workflow
- Southbound data collection system workflow
- Positioning and analytics service operation system workflow.

Sections 8.6.1.1–8.6.1.4 provide a short description for each workflow.

8.6.1.1   Service Activation

The proposed location-based analytics as a service platform relies on specific analytics service coordination as well as virtualized resources and functions orchestration capabilities which allow to introduce a high degree of automation in delivering and deploying analytics services on-demand. Specifically, the proposed location-based analytics as a service platform aims at deploying the various analytics services in the form of virtualized functions and services which are executed on top of hybrid virtualization infrastructures. The service activation system workflow is shown in Figure 8.7, which identifies the main involved system components and the interactions that are required to occur to have an automated deployment of analytics services within the platform. First, analytics services available for activation in the platform are described through specific templates and descriptors in the API catalogue, which can be accessed by the platform consumers (e.g. 3rd party users). The service activation workflow starts with an external platform consumer requesting the deployment of an analytics service through the dedicated APIs exposed by the service subscription module, which indeed allows to expose the content of the API catalogue and thus issue requests to create new instances of analytics services. The service subscription module verifies the credentials of the external platform consumer (through the access control module), and the request is forwarded to the analytics service coordinator. Here, the analytics service coordinator decomposes the analytics

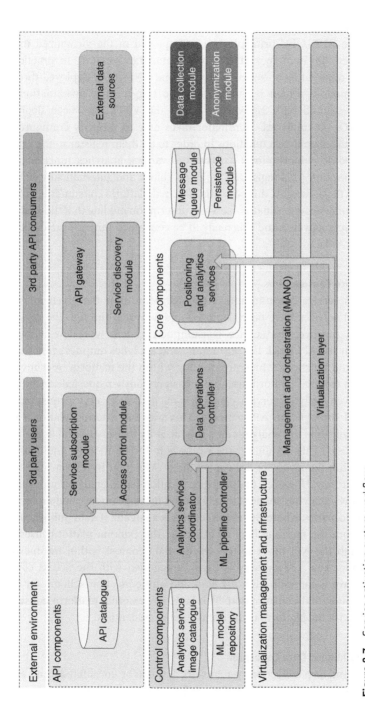

Figure 8.7 Service activation system workflow.

service activation request into one or more virtualized services and functions (following the ETSI NFV principles) to be deployed in the virtualized infrastructure to actually implement the required analytics services. In practice, it identifies the required NFV Network Services and VNFs to be deployed through the MANO module, which is then invoked to trigger their instantiation. As soon as the required NFV Network Services and VNFs have been deployed by the MANO, the analytics coordinator takes care to properly configure the various virtualized analytics functions in order to let them register to the service discovery module, which allows to have the various analytics functions and service consumption APIs registered in the API block and automatically exposed through the API gateway. At this point, the external platform consumer is notified by the chain analytics coordinator – service subscription module that the analytics service has been activated and it is ready to be consumed through the dedicated consumption APIs.

8.6.1.2   Service Consumption

The analytics service consumption is enabled soon after the given analytics functions and services have been deployed and activated through the previous workflow. In practice, the service consumption system workflow allows to consume from the API block the positioning and analytics outputs generated by the services and functions which are running within the platform. As shown in Figure 8.8, an external platform user, who has previously requested for the activation of a given analytics service, can access the dedicated service consumption APIs exposed through the API gateway. This way the external platform user can directly consume the analytics data produced by the service and apply it for its own purposes. This is valid for the case where the positioning and analytics functions within the given service expose direct APIs that have been registered in the API gateway. However, the proposed location-based analytics as a service platform allows analytics services to rely on the message queue module as well, for all those scenarios where the required data is a stream of data published and consumed from the message bus. In this case, the external platform user can access through the API block to the specific data context within the message queue module. The API gateway is tightly integrated with the access control module to have a common authorization and authentication scheme for both service activation and consumption. This way, any external platform consumer can interact with the platform using a single set of credentials.

8.6.1.3   Southbound Data Collection

The southbound data collection system workflow can be considered as an auxiliary workflow which allows the platform to be populated with data coming from

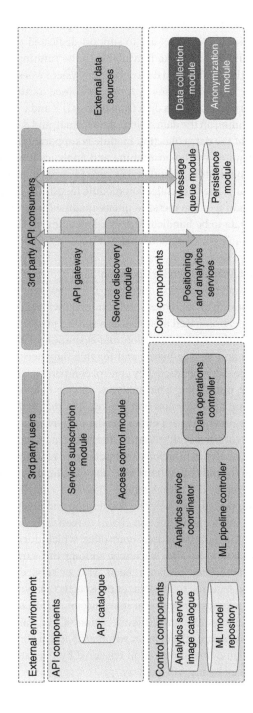

Figure 8.8 Service consumption system workflow.

heterogeneous sources. Indeed, for the execution and operation of the positioning and analytics functions, it is necessary to support dedicated mechanisms to ingest data into the platform and make it available to the running analytics services. Figure 8.9 shows the southbound data collection system workflow. The data collection module is the main data gateway for the proposed location-based analytics as a service platform, as it ingests data from both 3GPP and external sources into the platform. Depending on the data source, data format, and data context, different implementation of the data collection module is supported in the platform. In the case the data required by the positioning and analytics functions contain personal data, the anonymization module takes care to anonymize the data and make it privacy-compliant. The data collection module ingests the data into either the data persistence module or the message queue module (or both), depending on the specific type of data to be handled and also depending on the use that the related positioning and analytics functions make of it.

8.6.1.4 Positioning and Analytics Service Operation
The positioning and analytics service operation workflow regulates how the various functions composing an analytics service interact with the other components of the proposed location-based analytics as a service platform. Specifically, these interactions allow the actual execution and operation of the positioning and analytics functions and service logics, and for this they need to be assisted (in terms of control of the data operations) by control components of the platform. Figure 8.10 depicts the positioning and analytics service operation workflow. First, any positioning and analytics function deployed in the platform to realize a given analytics service, interacts with either the data persistence module or the message queue module (or even both) to satisfy their specific data requirements. Indeed, positioning and analytics functions are intended to read and write data in the persistence module (or publish/subscribe to any specific data context in the message queue) according to the actual service logic and analytics objective to be implemented and supported. The way the positioning and analytics functions interact each other (e.g. in a form of function chain) to realize the whole analytics service is regulated by the data operations controller, which during the service operation phase coordinates the data exchange among the various functions (e.g. by invoking dedicated internal data related APIs or service endpoints) to orchestrate the data flow required to implement the specific analytics service. For those positioning and analytics services that make use of AI/ML techniques, the ML pipeline controller allows to control the execution of ML training pipelines (in the virtualized infrastructure) either as an automated action part of the service action or as an explicit on-demand operation triggered by the external platform user through APIs exposed by the API gateway.

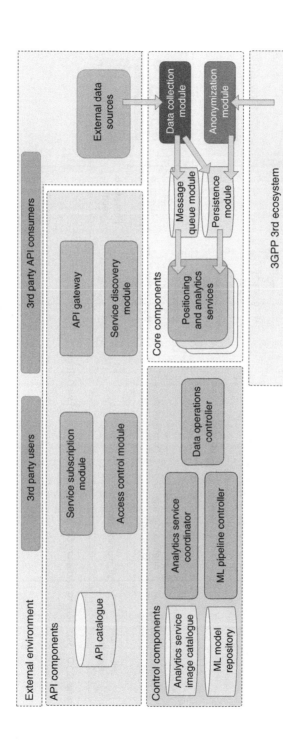

Figure 8.9 Southbound data collection system workflow.

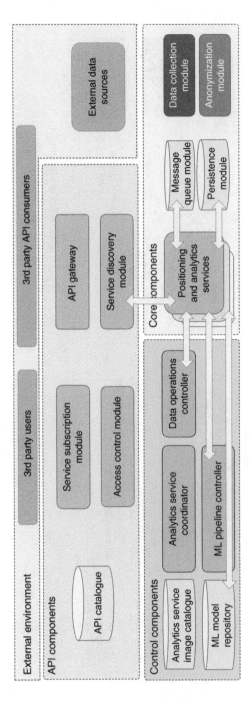

Figure 8.10 Positioning and analytics service operation system workflow.

8.6.2 Applicable Data Schema

For the implementation and integration of the various positioning and analytics functions and services within the proposed platform, as well as for the design and development of the platform components themselves, it is of crucial importance the utilization of interoperable, well-documented, and where possible existing opensource and industry-relevant standard data schema and formats. This allows to describe the data inputs and outputs at the various interface, and above all to foster the re-use and share of common positioning and analytics functions across several analytics services. This specific aspect, combined with the deployment and execution of the analytics services in the form of virtualized services and functions, substantially improves the replicability and scalability of the whole platform and the constituent functions. Sections 8.6.2.1–8.6.2.3 present few relevant opensource and industry-standard data schema and formats which are applicable for their use into the proposed location-based analytics as a service platform.

8.6.2.1 GeoJSON Data Format

The GeoJSON format [10] is a very widely used extension of the JSON format that is used to describe various geometries such as data points, lines (sequence of data points), and closed loops of points (polygons). It is also supporting multiple combinations of these primitives in order to be able to describe complex geometries that are found in geographical applications of the information and communication technology (ICT) domain. The proposed location-based analytics as a service platform is strongly related with the generation and consumption of geographical and geospatial data. As a consequence, a unified, standard, and well-documented data format is key for interoperability and accessibility of the analytics data produced and exposed by the platform. In particular, positioning and analytics functions can either consume or produce geometric shapes (e.g. for the identification of points/areas of interest or path identification functions), and are a good reference of services that can generate outputs in the GeoJSON format. In addition, the positioning functions running in the proposed platform can use the GeoJSON format to specify filtering boundary (area) for their localization computations. Another relevant attribute of the GeoJSON format is that a wide set of SQL implementations are available to translate GeoJSON format into native SQL functions commonly required by the analytics functions, making this format a relevant candidate native to model input/output data geospatial data for positioning and analytics functions.

8.6.2.2 JSON SQL Table Schema Format

The JSON SQL [11] table schema format is a widely used extension of the JSON format that is used to store self-contained SQL data rows. These objects are self-contained SQL rows and include both the payload (i.e. the main data of the row) and also the SQL schema that is used to store the data. Even though the data overhead of including both the schema and the payload information is considerable (especially for very complex data schema), the JSON SQL table schema format is a very helpful meta-data entry for any external or 3rd party data consumer that require to process the information. The use of the JSON SQL table schema format is applicable for all those data that require to be consumed by the external platform users. This means that it is suitable for being supported and implemented in either the data persistence or message queue module for those positioning and analytics data that is produced by analytics services and is suitable for external exposure.

8.6.2.3   3GPP Location Input Data

The positioning functions to be deployed and operated within the proposed location-based analytics as a service platform are specialized functions that translate raw 3GPP network measurement data into user and device positioning information. With the aim of achieving interoperability with the current data specifications of some 3GPP network components and functions, the inputs of the proposed positioning functions can be adapted to the specification provided by 3GPP TS 32.421 [12], describing trace concepts and requirements for the minimization of drive tests (MDT). These data include specific attribute related to the multicast-broadcast single-frequency network (MBSFN) such as MBSFN area identity, Carrier frequency, MBSFN reference signal received power (RSRP), MBSFN reference signal received quality (RSRQ), Multicast channel (MCH) block error rate (BLER) for signaling, MCH BLER for data, and related MCH index.

8.7   Platform Implementation: Available Technologies

The location-based analytics as a service platform described in this chapter is conceived to be implementable and deployable as a full working platform. Specifically, its native integration of hybrid virtualization technologies and infrastructures makes, beyond positioning and analytics functions and services, the platform itself implementable and deployable as a set on integrated microservices and cloud-native applications. As shown in Figure 8.11, several available opensource tools are suitable for adoption and integration to implement some of the components within the platform, as well as the positioning and analytics functions and services, thus reducing the platform development efforts. In the following, the list of applicable and relevant opensource tools and technologies is presented with a mapping to the system blocks and platform functionalities they can at least partially cover and implement.

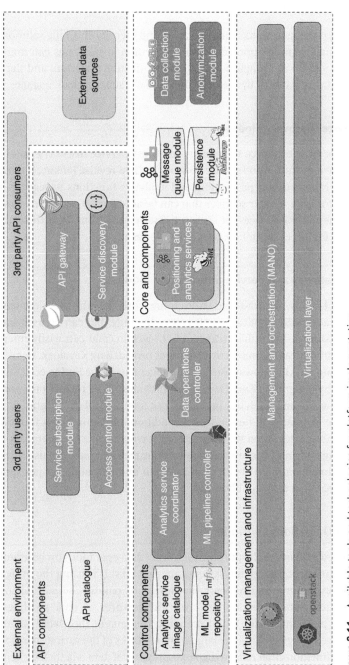

Figure 8.11 Available tools and technologies for platform implementation.

8.7.1 Access Control Module

Keycloak [13] is an opensource authentication and identity checking solution following the single sign-on approach. Java and python applications can integrate this 3rd party system as a single source of authentication (external) and the user plane metadata are handled by the local database of each Keycloak instance.

8.7.2 Service Discovery Module

Consul [14] is an opensource technology for REST service discovery functions, such as service registration and service network address reverse lookup. It provides a centralized repository for positioning and analytics functions to register their IP/port pairs for each of their services that can be accessed by generic consumers (i.e. other functions or services, or even the external platform users through the API gateway) for direct communication and interface.

8.7.3 API Gateway and Service Subscription Module

Spring Boot [15] is a Java-based web framework for the development of production-ready REST applications. It is a technology that can be used for the API gateway since it is flexible, scalable, light-weight, and can easily integrate with all the authentication and security features provided by Keycloak. In order to convert this into an API gateway, further enhancement of the Spring ecosystem is required, e.g. by integrating Zuul [16], which provides HTTP reverse proxy on the various positioning and analytics functions systems, as well as integration with Consul for the discovery of the services. In addition, the Spring Boot framework provides seamless integration with Swagger [17] which generates metadata from the positioning and analytics functions toward the external platform user that are essential for documenting APIs, in terms of input and output data types support and testing operation of the public APIs.

8.7.4 Data Operations Controller

Apache Airflow [18] is a multi-purpose high-level data operations orchestrator for a very wide variety of underlying systems and/or tasks. It is using the principles of directed acyclic graph (DAG) for the design of parallel, serial, dependent and independent jobs to build high-level architectures of data pipelines. In the proposed platform, it can be integrated to implement the Data Operations Controller, essentially invoking the large compute tasks of the various positioning and analytics functions that read/write large amounts of data (via their designated SQL or message bus interfaces). The sequence of invocation for these jobs is very crucial for their healthy operation, which is why for each analytics service is required to provide for executing its internal step-based tasks.

8.7.5 ML Pipeline Controller

Kubeflow [19] is a scalable ML platform that runs on Kubernetes and exploits all its capabilities to facilitate the development, deployment, and operations of virtualized ML solutions. It provides means to flexibly pre-process data and train various models in virtualized and containerized environments. Kubeflow can be used in the proposed platform to take care of the (virtualized) ML pipeline control operations, in a portable and scalable way. A pipeline, in Kubeflow, is a description of a ML workflow, including all of the steps in the workflow and how they combine in the form of a graph. A pipeline includes the definition of the inputs required to run the pipeline and the inputs and outputs of each step. It covers various steps in a ML pipeline, from training, to data pre-processing, data transformation, and serving.

8.7.6 ML Model Repository

MLFlow [20] is an opensource system designed to add dev-ops, deployment, and production readiness to the world of ML. It has capabilities such as model repository, experiment repository and orchestrates version control and meta-data storage to perform automated tasks that stabilize the delivery of updated models in modern production environments. The proposed location-based analytics as a service platform can integrate MLFLow as its ML model repository functionality, thus for storage of the trained models that are used by the positioning and analytics functions.

8.7.7 Data Collection Module

Apache Oozie [21] is an automated task scheduling tool that is commonly used in many big data cloud image providers. It uses a specialized XML-based data structure to define the types of jobs that it will execute and is used in conjunction with scripting languages and built-in client binaries. For the data collection module, the most appropriate client-side 3rd party software is the Apache Flume [22], which is a very widely used tool for the collection of external data source raw data and delivery at specified target locations. This is done by configuration files which specify the input and output sources.

8.7.8 Data Persistence Module

The technologies that can implement the data persistence functionalities in the platform are the de-facto standard for multi-virtual machine (VM) multi-disk, high availability, multi-node environments. Apache Hadoop distributed file system (HDFS) [23] can implement the storage layer at the lowest level. It is used in multiple cloud persistence deployments as it provides easy addition and removal of storage nodes (connected to virtual or actual disks) and it provides high throughput (HTTP-based) read and write operations. Apache Hive [24] is a technology that is built on top of Apache HDFS and is basically an SQL tabular data storage layer. It utilizes Hadoop [23] map-reduce binaries to transform various data sources into different file formats that are more optimized for compression or analytics. In addition, it gives the user access to an SQL compliant metadata storage that can allow for other query engines and SQL interfaces to perform SQL-like operations (which are subsequently converted into other types of operation on top of the HDFS nodes). Moreover, Trino [25] is an opensource distributed query engine that is optimized for multi-machine scaling on top of SQL data storage catalogues such as Apache Hive. By splitting the read and write operations between those two technologies, the platform can take advantage of the speed of Trino and the stability of Hive, map-reduced based transformation operations both operating on top of the HDFS-based file storage.

8.7.9 Message Queue

For the message queue module of the proposed platform, Apache Kafka [26] and RabbitMQ [27] are the production-ready tools which suit most the platform requirements. They both provide topic generation, multiple payload type formats such as JSON, XML, comma-separated value (CSV), Avro, and binary and they can be accessed easily by consumer and producer clients, which require very low code complexity but are very robust and performant, with specific mechanisms for parallel processing of data streams and integrity.

8.7.10 Virtualization layer

Kubernetes [28] and Openstack [29] can be integrated to realize the hybrid virtualization layer within the proposed platform. Openstack and Kubernetes are both de-facto standard virtualized infrastructure management solutions, and their combined use allows to deal with containerized and virtual machine-based applications. In addition, it enables a flexible and agile deployment of virtual functions and services at both edge (through Kubernetes) and cloud (through both Kubernetes and Openstack) compute locations. With this hybrid implementation approach, the platform virtualization layer also includes repositories for the virtual function software images, being them either virtual machines (by using the related Openstack image management service) or containers (by using legacy container registry solutions such those offered by Docker [30]).

8.7.11   Management and Orchestration

ETSI OSM [31] is an industry- and telco operator-led opensource initiative that provides a reference implementation of the ETSI NFV MANO architecture. It is basically the de-facto standard solution for managing and orchestrating VNFs and Network Services according to the ETSI NFV specifications. ETSI OSM implements various ETSI NFV standards, including northbound APIs for VNF and Network Service lifecycle management, and VNF and Network Service descriptors data model. It can be used to cover the MANO functionalities and system blocks, as it is natively capable to manage hybrid virtualization infrastructures where legacy VNFs and containerized network functions are deployed on top of Openstack and Kubernetes mixed infrastructures.

References

1 EU LOCUS Project. Deliverable 2.5, System Architecture: final version, 2021. URL https://www.locus-project.eu/wp-content/uploads/2021/12/D2.5_nbm_17-11-21.pdf.

2 B. Sayadi, N. Stasinopoulos, B. A. Jammal, T. Deiss, A. Ropodi, I. Fajjari, C. Patachia, D. Griffin, D. Breitgand, J. Martrat, R. Vilalta, S. Siddiqui, and G. Baldoni. From Webscale to Telco, the Cloud Native Journey, July 2018. URL https://doi.org/10.5281/zenodo.1306414.

3 C. Giebler, C. Gröger, E. Hoos, R. Eichler, H. Schwarz, and B. Mitschang. The data lake architecture framework: A foundation for building a comprehensive data lake architecture. In *Proceedings der 19. Fachtagung für Datenbanksysteme für Business, Technologie und Web (BTW 2021)*, Dresden, Germany, September 2021.

4 T. Zschörnig, R. Wehlitz, and B. Franczyk. A personal analytics platform for the Internet of Things - implementing Kappa Architecture with microservice-based stream processing. In *Proceedings of the 19th International Conference on Enterprise Information Systems - Volume 2: ICEIS*, pages 733–738, January 2017. doi: https://doi.org/10.5220/0006355407330738.

5 M. Kiran, P. Murphy, I. Monga, J. Dugan, and S. S. Baveja. Lambda architecture for cost-effective batch and speed big data processing. In *2015 IEEE International Conference on Big Data (Big Data)*, pages 2785–2792, 2015.

6 Apache Spark. https://spark.apache.org/, Last Accessed: May 5, 2023.

7 Seldon. https://www.seldon.io/, Last Accessed: May 5, 2023.

8 TS29.572. 5G System; Location Management Services; Stage 3, June 2019. Release 15.

9 TS33.107. 3rd Generation Partnership Project, (3GPP) 3G security; Lawful interception architecture and functions, June 2019. Release 15.

10 RFC7946. The GeoJSON Format, August 2016. RFC.

11 Data Protocols. JSON Table Schema, March 2015. Data Protocols.

12 TS32.421. Technical Specification, Trace concepts and requirements, December 2021. Release 16.

13 Keycloak. https://www.keycloak.org/, Last Accessed: October 31, 2022.

14 Consul. https://www.consul.io/, Last Accessed: October 31, 2022.

15 Spring Boot. https://spring.io/projects/spring-boot/, Last Accessed: October 31, 2022.

16 Zuul. https://github.com/Netflix/zuul, Last Accessed: October 31, 2022.

17 Swagger. https://swagger.io/, Last Accessed: October 31, 2022.

18 Apache Airflow. https://airflow.apache.org/, Last Accessed: October 31, 2022.

19 Kubeflow. https://www.kubeflow.org/, Last Accessed: October 31, 2022.

20 MLFlow. https://mlflow.org/, Last Accessed: October 31, 2022.

21 Apache Oozie. https://oozie.apache.org/, Last Accessed: October 31, 2022.

22 Apache Flume. https://flume.apache.org/, Last Accessed: October 31, 2022.

23 Apache Hadoop. https://hdfs.apache.org/, Last Accessed: October 31, 2022.

24 Apache Hive. https://hive.apache.org/, Last Accessed: October 31, 2022.

25 Apache Trino. https://trino.io/, Last Accessed: October 31, 2022.

26 Apache Kafka. https://kafka.apache.org/, Last Accessed: October 31, 2022.

27 RabbitMQ. https://www.rabbitmq.com/, Last Accessed: October 31, 2022.

28 Kubernetes. https://kubernetes.io/, Last Accessed: October 31, 2022.

29 Openstack. https://openstack.org/, Last Accessed: October 31, 2022.

30 Docker. https://www.docker.com/, Last Accessed: May 5, 2023.

31 ETSI OSM. https://osm.etsi.org/, Last Accessed: October 31, 2022.

9

Reference Standard Architectures

Giacomo Bernini[1], Aristotelis Margaris[2], Athina Ropodi[2] and Kostas Tsagkaris[2]

[1] *Nextworks, Pisa, Italy*
[2] *Incelligent P.C., Athens, Greece*

This chapter presents the reference standard architectures for the exposure of location-based analytics as a service. Table 9.1 lists the acronyms used in this Chapter.

9.1 Data Analytics in the 3GPP Architecture

Data analytics plays a critical role in the 3rd Generation Partnership Project (3GPP) view and specifications for the 5G network, and represents a key enabler for the generation of real-time operational intelligence to implement network automation. Therefore, 3GPP has introduced a general framework for data analytics in 5G infrastructures starting from the Rel. 15 and 16. In particular the 3GPP TR 23.791 defined a dedicated function within the 5G core, called Network Data Analytics Function (NWDAF) for the collection of data from the other network functions (and external data sources) and the delivery of data analytics services [1]. The NWDAF can operate in a centralized or distributed deployment model, with different levels of analytics granularity (e.g. global, per network slice). As depicted in Figure 9.1, it collects metrics or data analytics information locally elaborated from heterogeneous sources, like other 5G network functions, application functions, the management system as well as from external data repositories. All this data is processed by means of aggregation mechanisms, prediction algorithms, etc., with the aim of generating further analytics data to be then offered to other network functions, or if needed stored in dedicated data repositories. Starting from Rel. 16, the 3GPP TS 23.288 [2] specification is

Positioning and Location-based Analytics in 5G and Beyond, First Edition.
Edited by Stefania Bartoletti and Nicola Blefari Melazzi.

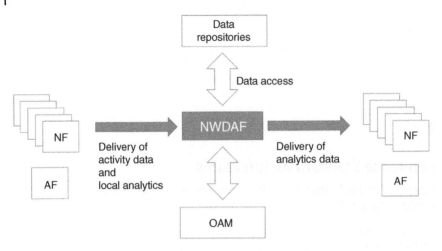

Figure 9.1 NWDAF in the framework for 5G network automation.

standardizing the NWDAF interfaces and the procedures enabling the consumption of data analytics services, supporting query/reply, as well as subscribe/notify models. In addition, the Open application programming interfaces (APIs) are also available in the 3GPP TS 29.520 specification [3]. The 3GPP TR 23.791 report [1] identifies several use cases showing the NWDAF applicability for different network control and management scenarios, with related decision-making logic spanning from the overall infrastructure down to the single network slice, service, or traffic flow. Examples of NWDAF related use cases include mobility management, quality of service (QoS) provisioning and adjustment, policy configuration, service-level agreement (SLA) guarantee assurance, allocation of edge resources, and load balancing. Depending on the specific use case, the target data include application and service-level monitoring data up to network key performance indicators (KPIs). However, the currently identified use cases and proposed solutions do not expose the requirement for UE positioning information for further contextualized, location-enriched analytics from which they would significantly benefit. The same stands, accordingly, for the NWDAF services (data types, semantics) specified for addressing these use cases and requirements.

9.1.1 Evolved Network Data Analytics in 3GPP R17

Beyond the current standard work on network data analytics, mostly defined in 3GPP Rel. 16, the NWDAF evolved in Rel. 17 in different directions to respond to various emerging data analytics requirements. Specifically, main novelties studied for Rel. 17 include: (i) Remaining key issues from Rel. 16, such as user equipment

Table 9.1 List of Acronyms.

Acronym	Definition
3GPP	3rd Generation Partnership Project
ADRF	Analytics data repository function
AEF	API exposing function
AI	Artificial intelligence
ANLF	Analytics logical function
API	Application programming interface
CAPIF	Common API framework
CCF	CAPI core function
DCCF	Data coordination and collection function
ETSI	European telecommunications standards institute
ISG	Industry specification group
KPI	Key performance indicator
MANO	Management and orchestration
MFAF	Messaging framework adaptor function
ML	Machine learning
MTLF	Model training logical function
NF	Network function
NFV	Network function virtualization
NWDAF	Network data analytics function
OSS	Operational support system
PLMN	Public land mobile network
PNF	Physical network function
POI	Point of interest
QoS	Quality of service
SBA	Service-based architecture
SEAL	Service enabler architecture layer for verticals
SLA	Service-level agreement
UE	User equipment
VIM	Virtualized infrastructure manager
VNF	Virtualized network function
VNFM	VNF manager
ZSM	Zero touch network and service management

(UE)-driven analytics and extensions of user data characteristics that can be used by NWDAF; (ii) NWDAF architecture enhancements toward multiple NWDAF instances in one Public Land Mobile Network (PLMN) including hierarchies, roles, and inter-NWDAF instance cooperation; (iii) NWDAF features enhancement, e.g. by enabling real-time or near real-time NWDAF communication, including mechanism for data collection and its optimization, e.g. in terms of load; and (iv) New use cases/key issues regarding interaction between NWDAF and machine learning (ML) model and training service owned by the operator. In particular, referring to the NWDAF use case on supporting dispersion analytics, according to which, in 3GPP Rel. 16, NWDAF can provide UE mobility statistics or prediction, mobility information is not sufficient to determine hot spots areas that may require additional operator attention, i.e. hot spots, points of interest (POIs), or trajectories on which users spend their data and thus need to be dynamically discovered through dispersion analytics. It is therefore identified as a key issue to check if the network functions (NFs) can benefit from the inclusion of such dispersion analytics, i.e. new type of analytics in addition to UE mobility and communication analytics.

Moreover, the NWDAF evolved along these lines in 3GPP TS 23.288 [2] where the data analytics framework is enhanced for Rel. 17. In particular, the NWDAF deployment model is shifting toward hierarchical and distributed architectures, introducing additional interfaces and procedures for inter-NWDAF communications. In this sense, the NWDAF evolved from a monolithic function toward a distributed functional entity with the composition of several elements which can interact and collaborate. A common approach, as depicted in Figure 9.2, is to split the NWDAF in distributed elements, with some functional entities located at the edge for real-time data collection and analytics, supported and

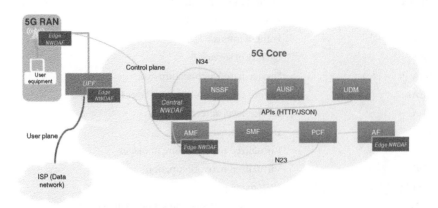

Figure 9.2 NWDAF distributed deployment model [4].

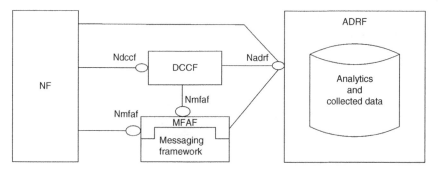

Figure 9.3 Architecture for network data analytics in Release 17 [2].

coordinated through a central NWDAF (e.g. deployed in the cloud) with functions of data aggregation, post-processing, including when required AI algorithms and training of ML models. This approach also allows for trained data model sharing between multiple NWDAF instances, with a solution based on an ML Model sharing architecture. In addition, the latest version of 3GPP TS 23.288 [2] defined a dedicated architecture for network data analytics collection, distribution, and storage. Here, additional functions are identified to facilitate the exchange and maintenance of data. As depicted in Figure 9.3, a Data Coordination and Collection Function (DCCF) coordinates the collection of data from multiple sources, their distribution and, if needed, their storage into the Analytics Data Repository Function (ADRF). Data delivery can be managed directly by this DCCF or leverage the Messaging Framework, e.g. when there are streaming of real-time data to distribute. Here, the Messaging Framework Adaptor Function (MFAF) enables the interaction between the involved 5G system components and the Messaging Framework offering data translation, formatting, and adaptation functionalities. The ADRF, on the other hand, stores the network analytics data and allows consumers to access historical data.

This architecture defined for Rel. 17 is aligned with the approach described in Chapter 8 on location-based analytics as a service basics, as data from multiple sources can be efficiently collected and distributed among several components and functions, with data consumers able to receive and exchange real-time streams of data (through a message bus/framework) or historical data (through a data persistency framework), depending on the specific analytics function scope and input/output data constraints. Moreover, the evolution of the network data analytics framework in 3GPP TS 23.288 [2] is also considering a functional split of the NWDAF into two functions, to clearly separate the pure analytics logic from the ML model training and management. This enables dedicated ML model and pipeline management functionalities (e.g. for training and creating new versions of the same model) to be clearly separated from the actual analytics functions

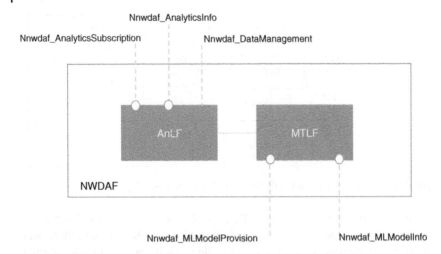

Figure 9.4 NWDAF split in ANLF and MTLF [2].

(pre-trained) ready to be deployed and operated for specific use cases. As shown in Figure 9.4, the Model Training Logical Function (MTLF) takes care of ML models training, exposing dedicated services and application programming interface (API)s for external functions to discover and query them. On the other hand, the Analytics Logical Function (ANLF) performs inference, elaborates statistics or predictions, and it can act as consumer of the MTLF service. A given NWDAF instance can contain MTLF, ANLF, or both, thus allowing for a full separation of model training and analytics functions.

9.1.2 Mapping with Location Data Analytics

To address the lack of contextualized and location-enriched analytics that, as introduced above, affect the NWDAF data types, semantics, and services, it is required to enrich them with important, meta-localization information/analytics. This would allow improving the existing NWDAF use cases and enable those identified in Chapters 6 and 7 for location analytics for verticals and location-aware network management. In practice, location analytics functions can be represented as specialized NWDAF, to undertake the part of data analytics exclusively associated with higher level, localization-awareness within the 5G Core. This opens two main options for these location analytics functions: (i) a tight integration with the existing NWDAF functionalities specifications, or (ii) a completely new entity in the 5G core. In the former case, the location-enriched NWDAF would inherit the integration with service-based paradigm and follow the (pub-sub, HTTP-based) interfaces description and procedures. Accordingly, this approach does not need to actually propose new interfaces, but can consider proposing analytics elements (i.e. analytics identifiers) and structures with regard to localization analytics. This could be done by adding application attributes and key performance indicator (KPI)s as the input/output data in some services described in 3GPP TS 23.288 [2]. In the latter case (new network function (NF) in the 5G core), proper interfaces (at least in terms of data and semantics) need to be specified in the context of the 5G Core, e.g. toward other network functions and NWDAF. Although this approach may appear ambitious, similar solutions can be found already in 3GPP Rel. 17, where completely new network functions are introduced for data collection, distribution, and storage for optimizing the 5G data architecture.

9.2 3GPP CAPIF

Introduced in Rel. 15, Common API Framework for 3GPP northbound application programming interfaces (APIs) (CAPIF) aims to offer a unified Northbound API framework across 3GPP network functions, ensuring a common, harmonized approach for their development and thus attempting to standardize to common aspects and relevant capabilities exposed through the northbound APIs [5–7]. It has since involved bringing – among others – features that support for third party domains and is currently on its Rel. 18 [8]. Several issues have been tackled within the Common API Framework (CAPIF). These issues involve a variety of processes, such as service publishing and discovery, service management, event handling (i.e. subscription, notifications), onboarding and offboarding functions, security aspects, as well as charging. In Figure 9.5, the functional model for CAPIF is presented. As shown, the main entities are:

- CAPIF core function (CCF): The CAPIF core function resides in the PLMN trust domain and supports APIs from both the PLMN trust domain and the third party trust domain. In this sense, it is a repository of service APIs, allowing for the discovery of the stored API by the API Invokers and API exposing functions (AEFs), authentication/authorization, and logging/charging all API invocations.
- API invoker: The applications invoking the services after discovering from the CCF the APIs and requesting authorization. It is typically a third party application provider who has a service agreement with PLMN operator and may reside within the same trust domain as the PLMN operator network.
- API exposing function (AEF): It provides the service to the API invoker and is in fact the entry point of the service API for the communication with the API invoker.
- API publishing function: It publishes the service API to the CCF for service discovery.

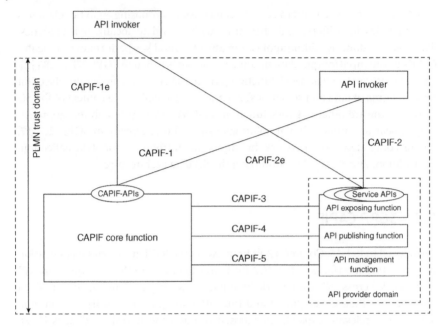

Figure 9.5 CAPIF architecture – functional model [8].

- API management function: This entity is related to managing the service APIs, i.e. auditing invocation logs from CCF, onboarding/offboarding API invokers, monitoring service APIs' status and events reported by the CCF, registering and maintaining registration information of the API provider domain functions, configuring the API provider policies.

Based on the aforementioned entities, it is easy to understand that the CAPIF architecture delivers a set of functional and system blocks that are essential to a framework for delivering northbound APIs. In essence, the API invokers play the roles of the third-party users/subscribers of a platform. Furthermore, it describes a "gateway" that may expose functions to the API invokers, while it includes service publish and discovery, service subscription and access control, monitoring, logging, and other capabilities.

9.3 3GPP SEAL

In order to enable vertical applications, the Service Enabler Architecture Layer for Verticals (SEAL) was introduced for Release 16 [9] and is currently on

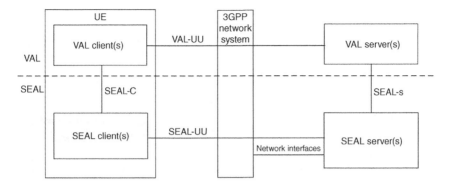

Figure 9.6 SEAL architecture – generic (on-network) functional model [10].

Release 18 [10]. As described, Service Enabler Architecture Layer for Verticals (SEAL)'s scope is to support vertical applications which require similar core capabilities in a timely manner. The specified application-enabling services can be reused across vertical applications and include the following:

- Location management;
- Group management;
- Configuration management;
- Identity management;
- Key management;
- Network resource management; and
- Data delivery.

The SEAL architecture describes a generic SEAL service functional model and specific SEAL service functional models that use the former as a reference. Indicatively, Figure 9.6 describes the generic on-network functional model, while the off-network functional model is similar but involves two UEs.

Whether on the UE or the server side the main entities are grouped in SEAL client(s) and SEAL server(s). More specifically, the former include client-side functionalities specific to the SEAL service, supporting Vertical Application Layer (VAL) client(s), while the latter include server-side functionalities. Lastly, it should also be noted that the SEAL server provide, APIs compliant with CAPIF as described in Section 9.2.

Specific to the location management service, SEAL aims to enable the sharing of location data between client-server, with varying granularity for vertical applications usage. Other requirements include on-demand or trigger-based location reporting, receiving updates on location information and sharing location information from 3GPP network systems to the vertical application.

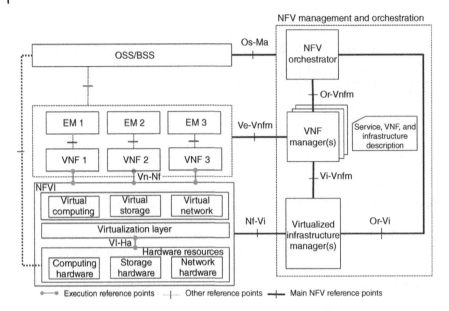

Figure 9.7 ETSI NFV architectural framework [11].

9.4 ETSI NFV

The European Telecommunication Standards Institute (ETSI) Network Function Virtualization (NFV) Industry Specification Group (ISG) has been the driver of the network transformation activities in ETSI, being at the core of the NFV technology definition and standardization, starting from the NFV term itself. As shown in the architecture diagram in Figure 9.7, the NFV scope is focused on the lifecycle of the Virtualized Network Functions (VNFs) and the services built by composing them. These may also include Physical Network Functions (PNFs), with common lifecycle management applied, irrespectivel of their nature, scope, technology. With NFV, network functions are implemented in software and can run on homogeneous, industry-standard commodity infrastructures. This software can then be moved to, or introduced in, different locations in the network as required. NFV simplifies the roll-out of network services, reduces deployment and operational costs, and facilitates network management automation. Up to now, the ISG has defined the initial framework and continued in the following phases working on interfaces, information models, and testing procedures. Detailed specifications for NFV-related workflows, data structures, and APIs are now under consolidation in the close-to-finish Release 4. At the current stage, the NFV community has produced a wide set of mature and stable specifications, has consolidated their applicability in the industry, and continues to work toward multi-vendor interoperability and addressing new architectural and application challenges. The available NFV standards of Releases 3 and 4 are already used in the industry to implement NFV products. NFV was originally conceived to help network service providers in the challenge of cost reduction and agility improvement, and it has resulted to be a key framework to enhance how services are requested and consumed by users. It is a necessary component for next-generation networks, and in particular for 5G. A central role in the ETSI NFV architecture is taken by the management and orchestration (MANO), which covers the lifecycle management of two entities: network services and VNF. As mentioned, VNF are network functions implemented in software and ready to be deployed and operated over virtualized infrastructures (supporting both traditional virtual machines and more cloud-native containerized applications), while the network service is the interconnection of VNF which chained together build a standalone service application logic. It is composed of three main building blocks: the NFV Orchestrator, VNF Manager (VNFM), and Virtualized Infrastructure Manager (VIM). VIM takes care to manage the virtualized infrastructure that could span across several locations (e.g. core data center locations and edge locations), and offer to the other MANO components specific APIs to create, update, and delete virtualized resources for running VNFs (e.g. virtual machines, containers, and virtual networking objects). The VNFM is responsible for the lifecycle management of each VNF building a network service, and coordinates all the phases for instantiation, scale, and termination, taking care of their performance and fault management as well. On top of the other components, the NFV Orchestrator coordinates the lifecycle of the network services, and is responsible to create them as the composition of multiple VNFs by interacting with the VNFM and the VIM. The ETSI NFV MANO enables to make automated the full lifecycle of network services and VNFs, starting from well-defined (and standardized within ETSI NFV) service and function description templates (i.e. the descriptors) and going through standard procedures and APIs for service instantiation, scale, performance monitoring, fault management. Therefore, the NFV software-enabled function and service lifecycle management provided by ETSI NFV MANO constitutes a new dimension to be considered in the network and service management field, and it has to be:

- Integrated with other end-to-end management tools and system, such as those for smart network management within advanced Operations Support System (OSS), as well as network slicing in 5G scenarios
- Considered as an essential enabling technology for network automation and simplification, with services to be easily deployed in multiple instances in distributed locations

- Considered as a key target for new operational architectures and deployments where automation of virtualized environment and the services running on top is required.

9.4.1 Mapping with Location Analytics Functions Management

According to the above, the ETSI NFV framework, and in particular the MANO architecture becomes very relevant for the Location-based Analytics as a Service platform introduced in Chapter 8, as it offers the APIs, the data models, and the workflow logics for building complex services as the combination of individual virtualized functions running in software and deployed over a virtualized infrastructure. In particular, localization analytics services can be implemented as a specific case of NFV network services, built by the combination of data collection and storage functions, localization functions, analytics functions, machine learning functions to be interconnected according to their data requirements at both input and output sides. Therefore, according to the specific needs in terms of required data, localization scope, localization features, dedicated combinations of localization-related functions (including analytics and machine learning) can be prepared to compose proper localization services. This means that the NFV network service descriptor should be designed, prepared, and onboarded in the MANO system according to the requirements of the involved functions. The ETSI NFV MANO can be then considered as a reference approach for managing and coordinating the lifecycle of the localization analytics services, enabling full automation in their deployment and operation over virtualized infrastructures.

9.5 ETSI Zero Touch Network and Service Management (ZSM)

The ETSI Zero touch network and Service Management (ZSM) ISG is aiming at defining a new, horizontal, and vertical end-to-end operable framework. It enables agile, efficient, and qualitative management and automation of emerging and future networks and services. ZSM targets a network and service management architecture where all operational processes and tasks (e.g. delivery, deployment, configuration, assurance, and optimization) are executed automatically. Ideally, this is done with full automation in multi-vendor environments. The existing work from ETSI NFV solves dedicated aspects of network and service management, and has defined management capabilities for their respective network function and service virtualization domain. On top of this, ETSI ZSM aims at providing a holistic service management concept which, among others, enables the integration of NFV and edge computing management demands with other relevant service management aspects such as data collection, analytics, intelligence. However, NFV-like architectures are still based on monolithic control and orchestration solutions, mostly dedicated to the management of the pipelined network services, where the lack of agility in the service lifecycle and operation is still a clear limitation, especially when it comes to fulfilling heterogeneous service constraints posed by 5G services and beyond. ZSM goes in the direction of overcoming such limitations, and its reference architecture is built around a set of building blocks that collectively enable construction of more complex management services and management functions. The clear identification and separation of management domains provide means to isolate management duties (possibly referring to very heterogeneous technologies), considering boundaries of different nature (technological, administrative, geographical, etc.). Each management domain provides a set of ZSM management services, realized by functions that expose and/or consume a set of end-points. An end-to-end service management domain is a special management domain responsible for the cross-domain management and coordination, which glue all of the single domain management services, functions, and end-points. At the core of the ZSM architecture there are the domain integration fabrics and the cross-domain integration fabric, which facilitate the provision of services and the access to them through the related end-points across the various domains. It also includes specific services for the communication between management functions, which enable the exchange of management data to consumers. In addition, dedicated domain data services provide the mean to persist data and access it still through the integration fabrics. According to ZSM principles, management services can be logically grouped according to the functionality offered (such as data collection, analytics, intelligence, orchestration, control). Figure 9.8 depicts the ZSM framework reference architecture. The possibility to flexibly compose management services together with the exchange of management data provide the foundation of an innovative agile service management that makes easier the integration of the various management aspects (from data collection to orchestration and analysis) enabling the closure of the control loop through network and service optimization processes.

9.5.1 Mapping with Location Analytics Services Management

Based on all of the above, the ETSI ZSM approach and principles fit with the requirements and needs of the Location-based Analytics as a Service platform introduced in Chapter 8, for what concerns an agile framework for composing different localization services under a common agile and flexible architecture, where the integration fabric can drive the cooperation of orchestration, data

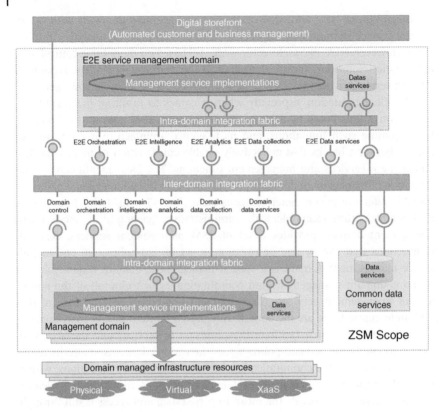

Figure 9.8 ETSI ZSM reference architecture [12].

collection and storage, analytics services following a service-based architecture (SBA) approach with decoupled services and functions. In practice, ETSI ZSM provides the means to enhance the ETSI NFV MANO capabilities (purely dedicated to orchestration and lifecycle management of standalone network services). This enables a more comprehensive approach where other crucial aspects of service operation are considered and tightly integrated, as data analytics and intelligence, which can then be exposed to external entities to build applications on top. This is crucial for properly enabling the support of location analytics for verticals and location-aware network management use cases on top of localization services and functions. In practice, the Location-based Analytics as a Service platform is conceptually aligned to the ETSI ZSM architecture and principles, as data exposure and data analytics services as defined in ETSI ZSM 002 [12] are in line with the capability to offer localization analytics and localization data as services to the application layer.

References

1 TR23.791. 3rd Generation Partnership Project (3GPP) Technical Report; study of enablers for network automation for 5G, 2019.

2 TR23.288. 3rd Generation Partnership Project (3GPP) Technical Specification, architecture enhancements for 5G system (5GS) to support network data analytics services, December 2021. Release 17.

3 TS29.520. 3rd Generation Partnership Project (3GPP) Technical Specification, 5G system; network data analytics services; Stage 3, December 2021. Release 17.

4 TM Forum. NWDAF: automating the 5G network with machine learning and data analytics, June 2020.

5 TS23.222. 3rd Generation Partnership Project, (3GPP) Technical Specification, Functional architecture and information flows to support Common API Framework for 3GPP northbound APIs; Stage 2, October 2017. Release 15.

6 TS33.122. 3rd Generation Partnership Project, (3GPP) Technical Specification, Security Aspects of Common API Framework for 3GPP Northbound APIs, October 2018. Release 15.

7 TS29.222. 3rd Generation Partnership Project, (3GPP) Technical Specification, Common API Framework for 3GPP Northbound APIs; Stage 3, March 2018. Release 15.

8 TS23.222. 3rd Generation Partnership Project (3GPP) Technical Specification, Functional architecture and information flows to support Common API Framework for 3GPP Northbound APIs; Stage 2, March 2023. Release 18.

9 TS23.434. 3rd Generation Partnership Project (3GPP) Technical Specification, Service Enabler Architecture Layer for Verticals; Functional architecture and information flows, January 2019. Release 16.

10 TS23.434. Technical Specification, Service Enabler Architecture Layer for Verticals (SEAL); Functional architecture and information flows, April 2023. Release 18.

11 GS NFV 002 v1.2.1. Network Functions Virtualisation (NFV); architectural framework, December 2014.

12 ETSI ZSM, GS ZSM 002 v1.1.1. Zero-touch network and service management (ZSM); reference architecture, August 2019.

Index

Positioning and Location-based Analytics in 5G and Beyond, First Edition.
Edited by Stefania Bartoletti and Nicola Blefari Melazzi.
© 2024 The Institute of Electrical and Electronics Engineers, Inc. Published 2024 by John Wiley & Sons, Inc.

Printed and bound by CPI Group (UK) Ltd, Croydon, CR0 4YY

23/10/2023

08154975-0001